I.M. Gel'fand   E.G. Glagoleva   A.A. Kirillov

The Method of Coordinates

1990

Birkhäuser
Boston • Basel • Berlin

I.M. Gel'fand
University of Moscow
117234 Moscow
USSR

E.G. Glagoleva
University of Moscow
117234 Moscow
USSR

A.A. Kirillov
University of Moscow
117234 Moscow
USSR

The Method of Coordinates was originally published in 1966 in the Russian language under the title Metod koordinat.

The english language edition was translated and adapted by Leslie Cohn and David Sookne under the auspices of the Survey of Recent East European Mathematical Literature, conducted by the University of Chicago under a grant from the National Science Foundation. It is republished here with permission from the University of Chicago for its content and the MIT Press for its form.

The Library of Congress Cataloging-in-Publication Data
Gel'fand, I. M. (Izrail ' Moiseevich)
    [Metod koordinat. English]
    The method of coordinates / I.M. Gel'fand, E.G. Glagoleva, and
A.A. Kirillov ' translated and adapted from the Russian by Leslie
Cohn and David Sookne.
        p. cm.
    Translation and adaptation of: Metod koordinat.
    Reprint. Originally published: Cambridge: M.I.T. Press,
1967. (Library of school mathematics; v. 1)
    ISBN 0-8176-3533-5
    1. Coordinates. I. Glagoleva, E. G. (Elena Georgievna)
II. Kirillov, A. A. (Aleksandr Aleksandrovich), 1936- .
III. Title.
QA556.G273 1990   90-48209
516'.16--dc20   CIP

ISBN 0-8176-3533-5
ISBN 3-7643-3533-5

Text reproduced with permission of the MIT Press, Cambridge, Massachusetts, from their edition published in 1967.
Printed and bound by R.R. Donnelley and Sons, Harrisonburg, Virginia.
Printed in the U.S.A.

9 8 7 6 5 4 3 2 1

Preface

Dear Students,

We are going to publish a series of books for high school students. These books will cover the basics in mathematics. We will begin with algebra, geometry and calculus. In this series we will also include two books which were written 25 years ago for the Mathematical School by Correspondence in the Soviet Union. At that time I had organized this school and I continue to direct it.

These books were quite popular and hundreds of thousands of each were sold. Probably the reason for their success was that they were useful for independent study, having been intended to reach students who lived in remote places of the Soviet Union where there were often very few teachers in mathematics.

I would like to tell you a little bit about the Mathematical School by Correspondence. The Soviet Union, you realize, is a large country and there are simply not enough teachers throughout the country who can show all the students how wonderful, how simple and how beautiful the subject of mathematics is. The fact is that everywhere, in every country and in every part of a country there are students interested in mathematics. Realizing this, we organized the School by Correspondence so that students from 12 to 17 years of age from any place could study. Since the number of students we could take in had to be restricted to about 1000, we chose to enroll those who lived outside of such big cities as Moscow, Leningrad and Kiev and who inhabited small cities and villages in remote areas. The books were written for them. They, in turn, read them, did the problems and sent us their solutions. We never graded their work -- it was forbidden by our rules. If anyone was unable to solve a problem then some personal help was given so that the student could complete the work.

Of course, it was not our intention that all these students who studied from these books or even completed the School should choose mathematics as their future career. Nevertheless, no matter what they would later choose, the results of this training remained with them. For many, this had been their first experience in being able to do something on their own -- completely independently.

I would like to make one comment here. Some of my American colleagues have explained to me that American students are not really accustomed to thinking and working hard, and for this reason we must make the material as attractive as possible. Permit me to not completely agree with this opinion. From my long experience with young students all over the world I know that they are curious and inquisitive and I believe that if they have some clear material presented in a simple form, they will prefer this to all artificial means of attracting their attention -- much as one buys books for their content and not for their dazzling jacket designs that engage only for the moment.

The most important thing a student can get from the study of mathematics is the attainment of a higher intellectual level. In this light I would like to point out as an example the famous American physicist and teacher Richard Feynman who succeeded in writing both his popular books and scientific works in a simple and attractive manner.

I.M. Gel'fand

Foreword

The Method of Coordinates is the method of transferring a geometrical image into formulas, while in the previous book *Functions and Graphs* you learned how to transfer formulas into picutres.

The systematic development of this method was proposed by the outstanding French philosopher and mathematician René Descartes about 350 years ago. It was a great discovery and very much influenced the development not only of mathematics but of other sciences as well. Even today you cannot avoid the method of coordinates. In any image on the computer or TV, every transmission of the picture from one place to another uses the transformation of the visual information into numbers -- and vice versa.

Note to Teachers

This series of books includes the following material:

1. *Functions and Graphs*
2. *The Method of Coordinates*
3. *Algebra*
4. *Geometry*
5. *Calculus*
6. *Combinatorics*

Of course, all of the books may be studied independently. We would be very grateful for your comments and suggestions. They are especially valuable because books 3 through 6 are in progress and we can incorporate your remarks. For the book *Functions and Graphs* we plan to write a second part in which we will consider other functions and their graphs, such as cubic polynomials, irrational functions, exponential function, trigonometrical functions and even logarithms and equations.

Contents

Chapter 2   Four-Dimensional Space   54

Chapter 3   The Four-Dimensional Cube   62

Introduction

When you read in the newspapers of the launching of a new space satellite, pay attention to the statement: "The satellite was placed in an orbit close to the one that was calculated." Consider the following problem: How can we calculate — that is, study numerically — the orbit of the satellite — a line? For this we must be able to translate geometrical concepts into the language of numbers and in turn be able to define the position of a point in space (or in the plane, or on the surface of the earth, and so on) with the aid of numbers.

The method of coordinates is the method that enables us to define the position of a point or of a body by numbers or other symbols.

The numbers with which the position of the points is defined are called the *coordinates* of the point.

The geographical coordinates (with which you are familiar) define the position of a point on a surface (the surface of the earth); each point on the surface of the earth has two coordinates: latitude and longitude.

In order to define the position of a point in space, we need not two numbers but three. For example, to define the position of a satellite, we can indicate its height above the surface of the earth, and also the latitude and longitude of the point which it is over.

If the trajectory of the satellite is known — that is, if we know the line along which it is moving — then in order to define the position of the satellite on this line, it is enough to indicate one number, for example, the distance traveled by the satellite from some point on the trajectory.[1]

Similarly, the method of coordinates is used for defining the position of a point on a railroad track: one shows the number of kilometer posts. This number then is the coordinate of the point on the railway line. In the name "The Forty-second Kilometer Platform," for example, the number 42 is the coordinate of the station.

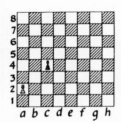

Fig. 1

A peculiar kind of coordinates is used in chess, where the position of figures on the board is defined by letters and numbers. The vertical columns of squares are indicated by letters of the alphabet and the horizontal rows by numbers. To each square on the board there correspond a letter, showing the vertical column in which the block lies, and a number, indicating the row. In Fig. 1 the white pawn lies in square $a2$ and the black one in $c4$. Thus, we can regard $a2$ as the coordinates of the white pawn and $c4$ as those of the black.

The use of coordinates in chess allows us to play the game by letter. In order to announce a move, there is no need to draw the board and the positions of the figures. It is sufficient, for example, to say: "The Grand Master played $e2$ to $e4$," and everyone will know how the game opened.

The coordinates used in mathematics allow us to define numerically the position of an arbitrary point in space, in a plane, or on a line. This enables us to "cipher" various kinds of figures and to write them down with the aid of numbers. You will find one of the

[1]Sometimes we say that a line has one dimension, a surface, two, and space, three. By the dimension, then, we mean the number of coordinates defining the position of a point.

examples of this kind of ciphering in Exercise 1 in Section 4.

The method of coordinates is particularly important because it permits the use of modern computers not only for various kinds of computations but also for the solution of geometrical problems, for the investigation of arbitrary geometrical objects and relations.

We shall begin our acquaintance with the coordinates used in mathematics with an analysis of the simplest case: defining the position of a point on a straight line.

PART I

CHAPTER 1

The Coordinates of Points on a Line

1. The Number Axis

In order to give the position of a point on a line, we proceed in the following manner. On the line we choose an *origin* (some point O), a *unit of measurement* (a line segment e), and a *direction* to be considered positive (shown in Fig. 2 by an arrow).

Fig. 2

A line on which an origin, a unit of measurement, and a positive direction are given will be called a *number axis*.

To define the position of a point on a number axis it suffices to specify a single number — $+5$, for example. This will indicate that the point lies 5 units of measurement from the origin in the positive direction.

The number defining the position of a point on a number axis is called the *coordinate* of the point on this axis.

The coordinate of a point on a number axis is equal to the distance of the point from the origin of coordinates expressed in the chosen units of measurement and taken with a plus sign if the point lies in the positive direction from the origin, and with a minus sign in the opposite case. The origin is frequently called the *origin of coordinates*. The coordinate of the origin (the point O) is equal to zero.

7

We use the designation: $M(-7)$, $A(x)$, and so on. The first of these indicates the point M with the coordinate -7; the latter, the point A with the coordinate x. Frequently we say more briefly: "the point minus seven," "the point x," and so on.

In introducing coordinates, we have set up a correspondence between numbers and points on a straight line. In this situation the following remarkable property is satisfied: to each point of the line there corresponds one and only one number, and to each number there corresponds one and only one point on the line.

Let us introduce a special term: a correspondence between two sets is said to be *one-to-one* if for each element of the first set there is one element of the second set and (in this same correspondence) each element of the second set

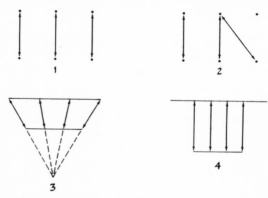

corresponds to some element of the first set. In Examples 1 and 3 in the figure the correspondence is one-to-one, but in 2 and 4 it is not. At first glance it appears quite simple to set up a one-to-one correspondence between the points on a line and the numbers. However, when mathematicians considered the matter, it turned out that, in order to elucidate the exact meaning of the words in this statement, a long and complicated theory had to be created. For immediately two "simple" questions arise which are difficult to answer: What is a number and what does one mean by a point?

These questions are related to the foundations of geometry and to the axiomatics of numbers. We shall examine the latter somewhat more closely in another booklet in our series.

Although the question of defining the position of a point on a line is quite simple, we must examine it carefully in order to become accustomed to seeing geometrical relations in numerical ones, and vice versa.

Test yourself.

If you have correctly understood Section 1, you will have no difficulty with the exercises we have prepared for you. If you cannot do them, this means that you have left out or not understood something. In that case, go back and read the passage over.

EXERCISES

1. (*a*) Draw on a number axis the points:

$$A(-2), \quad B\left(\tfrac{13}{3}\right), \quad K(0).$$

(*b*) On a number axis draw the point $M(2)$. Find the two points A and B on the number axis located a distance of three units from the point M. What are the coordinates of the points A and B?

2. (*a*) It is known that the point $A(a)$ lies to the right of the point $B(b)$.[1] Which number is greater: a or b?

(*b*) Without drawing the points on a number axis, decide which of the two points is to the right of the other: $A(-3)$ or $B(-4)$, $A(3)$ or $B(4)$, $A(-3)$ or $B(4)$, $A(3)$ or $B(-4)$.

3. Which of these two points lies to the right of the other: $A(a)$ or $B(-a)$? (**Answer.** We cannot say. If a is positive, then A lies to the right of B; if a is negative, then B lies to the right of A.)

4. Consider which of the following points lies to the right of the other: (*a*) $M(x)$ or $N(2x)$; (*b*) $A(c)$ or

---

[1]From here on we shall suppose that the axis is drawn horizontally and that the positive direction is from left to right.

$B(c + 2)$; (c) $A(x)$ or $B(x - a)$. (**Answer.** If a is greater than zero, then A is to the right; if a is less than zero, then B is to the right. If $a = 0$, then A and B coincide.) (d) $A(x)$ or $B(x^2)$.

5. Draw the points $A(-5)$ and $B(7)$ on a number axis. Find the coordinate of the center of the segment AB.

6. With a red pencil, mark off on a number axis the points whose coordinates are: (a) whole numbers; (b) positive numbers.

7. Mark off all the points x on a number axis for which: (a) $x < 2$; (b) $x \geq 5$; (c) $2 < x < 5$; (d) $-3\frac{1}{4} \leq x \leq 0$.

2. The Absolute Value of a Number

By the *absolute value* of the number x (or the *modulus* of the number x) we mean the distance of the point $A(x)$ from the origin of coordinates.

The modulus of the number x is denoted by vertical lines: $|x|$ is the modulus of x.

For example, $|-3| = 3$, $|\frac{1}{2}| = \frac{1}{2}$.

From this it follows that

$$\text{if } x > 0, \text{ then } |x| = x,$$
$$\text{if } x < 0, \text{ then } |x| = -x,$$
$$\text{if } x = 0, \text{ then } |x| = 0.$$

Since the points a and $-a$ are located at the same distance from the origin of coordinates, the numbers a and $-a$ have the same absolute value: $|a| = |-a|$.

EXERCISES

1. What values can the expression $|x|/x$ take on?

2. How can the following expressions be written without using the absolute value sign: (a) $|a^2|$; (b) $|a - b|$, if $a > b$; (c) $|a - b|$, if $a < b$; (d) $|-a|$, if a is a negative number?

10

3. It is known that $|x - 3| = x - 3$. What can x be?

4. Where on a number axis can the point x lie if (a) $|x| = 2$; (b) $|x| > 3$?

Solution. If x is a positive number, then $|x| = x$, and so $x > 3$; if x is a negative number, then $|x| = -x$; thus from the inequality $-x > 3$, it follows that $x < -3$. (**Answer.** To the left of the point -3 or to the right of the point 3. This answer can be gotten more quickly if one takes into account that $|x|$ is the distance of the point x from the origin of coordinates. It is then clear that the desired points are located at a distance from the origin which is greater than 3. The answer is obtained from a sketch.) (c) $|x| \leq 5$; (d) $3 < |x| < 5$? (e) Show where the points lie for which $|x - 2| = 2 - x$.

5. Solve the equations: (a) $|x - 2| = 3$; (b) $|x + 1| + |x + 2| = 1$. (**Answer.** The equation has infinitely many solutions: the collection of all solutions fills the segment $-2 \leq x \leq -1$; that is, any number which is greater than or equal to -2 and less than or equal to -1 satisfies the equation.)

3. The Distance Between Two Points

Let us begin with an exercise. Find the distance between the points:

(a) $A(-7)$ and $B(-2)$;

(b) $A(-3\frac{1}{2})$ and $B(-9)$.

It is not difficult to solve these problems since, knowing the coordinates of the points, one can figure out which is to the right of the other, how they are situated with respect to the origin of coordinates, and so on. After this it is quite easy to see how to calculate the desired distance.

We now propose that you derive a general formula for the distance between two points on a number axis, that is, that you solve the following problem:

Problem. Given the points $A(x_1)$ and $B(x_2)$; define the distance $d(A, B)$ between these points.[1]

Solution. Since the concrete values of the co-ordinates of the points are not known, it is necessary to draw all possible cases of the mutual relation of the points A, B, and C (the origin).

There are six such cases. Let us first examine the three cases in which B is to the right of A (Fig. 3a, b, and c).

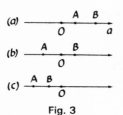

Fig. 3

In the first of these (Fig. 3a) the distance $d(A, B)$ is equal to the difference of the distances of the points B and A from the origin. Since in this case x_1 and x_2 are positive,

$$d(A, B) = x_2 - x_1.$$

In the second case (Fig. 3b) the distance is equal to the sum of the distances of the points B and A from the origin; that is, as before,

$$d(A, B) = x_2 - x_1,$$

since in this case x_2 is positive and x_1 negative.

Show that in the third case (Fig. 3c) the distance will be defined by the same formula.

The other three cases (Fig. 4) differ from those already considered in that the roles of the points A and B have been interchanged. In each of these cases one can check that the distance between the points A and B is equal to

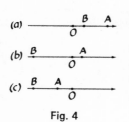

Fig. 4

$$d(A, B) = x_1 - x_2.$$

Thus in all cases where $x_2 > x_1$, the distance $d(A, B)$ is equal to $x_2 - x_1$, and in all cases where $x_1 > x_2$ this distance is equal to $x_1 - x_2$. Recalling the definition of the absolute value, one can write this

[1]The letter d is usually used for designating a distance. The expression $d(A, B)$ designates the distance between the points A and B.

12

using a single formula valid in all six cases:

$$d(A, B) = |x_1 - x_2|.$$

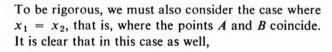

If desired this formula can also be written as

$$d(A, B) = |x_2 - x_1|.$$

To be rigorous, we must also consider the case where $x_1 = x_2$, that is, where the points A and B coincide. It is clear that in this case as well,

$$d(A, B) = |x_2 - x_1|.$$

Thus the problem we have set has been solved in full.

EXERCISES

1. Mark off on a number axis the points x for which (a) $d(x, 7) < 3$; (b) $|x - 2| > 1$; (c) $|x + 3| = 3$.

2. On a number axis two points $A(x_1)$ and $B(x_2)$ are given. Find the coordinate of the center of the segment AB. (**Hint.** In solving this problem you must examine all possible cases of the positions of A and B on the number axis or else write down a solution which would be valid at once for all cases.)

3. Find the coordinate of the point on the number axis which is located twice as close to the point $A(-9)$ as to the point $B(-3)$.

4. Solve the equations (a) and (b) of Exercise 5 on page 9 using the concept of the distance between two points.

5. Solve the following equations:

    (a) $|x + 3| + |x - 1| = 5$;

    (b) $|x + 3| + |x - 1| = 4$;

    (c) $|x + 3| + |x - 1| = 3$;

    (d) $|x + 3| - |x - 1| = 5$;

    (e) $|x + 3| - |x - 1| = 4$;

    (f) $|x + 3| - |x - 1| = 3$.

CHAPTER 2

The Coordinates of Points in the Plane

4. The Coordinate Plane

In order to define the coordinates of a point in the plane, we shall draw two mutually perpendicular number axes. One of these will be called the *abscissa* or *x-axis* (or *Ox*) and the other the *ordinate* or *y-axis* (or *Oy*).

The direction of the axes is usually chosen so that the positive semiaxis *Ox* will coincide with the positive semiaxis *Oy* after a 90° rotation counterclockwise (Fig. 5). The point of intersection of the axes is called the *origin of coordinates* (or simply *origin*) and is designated by the letter *O*. It is taken to be the origin of coordinates for each of the number axes *Ox* and *Oy*. The units of measurement on these axes are chosen, as a rule, to be identical.

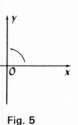

Fig. 5

Let us take some point *M* on the plane and drop perpendiculars from it to the axis *Ox* and to the axis *Oy* (Fig. 6). The points of intersection M_1 and M_2 of these perpendiculars with the axes are called the *projections* of the point *M* on the coordinate axes.

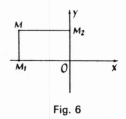

Fig. 6

The point M_1 lies on the coordinate axis *Ox*, and so there is a definite number *x* corresponding to it. This number is taken to be the coordinate of *M* on the

x-axis. In the same way the point M_2 corresponds to some number y — its coordinate on the y-axis.

In this way, to each point M lying in the plane there correspond two numbers x and y, which are called the *rectangular Cartesian coordinates* of the point M. The number x is called the *abscissa* of the point M, and y is its *ordinate*.

On the other hand, for each pair of numbers x and y it is possible to determine a point in the plane for which x is the abscissa and y the ordinate.

Now we have set up a one-to-one correspondence[1] between the points in the plane and pairs of number x and y taken in a definite order (first x, then y).

Thus, the *rectangular Cartesian coordinates* of a point in the plane are called the coordinates on the coordinate axes of the projections of the point on these axes.

The coordinates of the point M are usually written in the following manner: $M(x, y)$. The abscissa is written first and then the ordinate. Sometimes instead of "the point with the coordinates $(3, -8)$" one speaks of "the point $(3, -8)$."

The coordinate axes divide the plane into four *quarters* (*quadrants*). The first quadrant is taken to be the quadrant between the positive semiaxis Ox and the positive semiaxis Oy. The other quadrants are numbered consecutively counterclockwise (Fig. 7).

To master the notion of coordinates in the plane, do a few exercises.

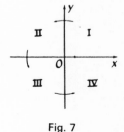

Fig. 7

EXERCISES

First we provide some quite simple problems.

1. What do the following symbols mean?

$(6, 2)$, $(9, 2)$, $(12, 1)$, $(12, 0)$, $(11, -2)$, $(9, -2)$,

---

[1]A one-to-one correspondence between the points of a plane and pairs of numbers is a correspondence such that to each point there corresponds one definite pair of numbers and to each pair of numbers there corresponds one definite point (cf. p. 8).

(4, −2), (2, −1), (1, 1), (−1, 1), (−2, 0), (−2, −2), (2, 1), (5, 2), (12, 2), (9, 1), (10, −2), (10, 0), (4, 1), (2, 2), (−2, 2), (−2, 1), (−2, −1), (0, 0), (2, 0), (2, −2), (4, −1), (12, −1), (12, −2), (11, 0), (7, 2), (4, 0), (9, 0), (4, 2).

2. Without drawing the point $A(1, −3)$, say what quadrant it lies in.

3. In which quadrants can a point be located if its abscissa is positive?

4. What will be the signs of the coordinates of points located in the second quadrant? In the third quadrant? In the fourth?

5. A point is taken on the Ox axis with the coordinate −5. What will its coordinates be in the plane? (**Answer.** The abscissa of the point is equal to −5, and the ordinate is equal to zero.)

Here are some more complicated problems.

6. Draw the points $A(4, 1)$, $B(3, 5)$, $C(−1, 4)$, and $D(0, 0)$. If you have drawn them correctly, you have the vertices of a square. What is the length of the sides of this square? What is its area?[1] Find the coordinates of the midpoints of the sides of the square. Can you show that $ABCD$ is a square? Find four other points (give their coordinates) that form the vertices of a square.

7. Draw a regular hexagon $ABCDEF$. Take the point A as the origin; direct the abscissa axis from A to B; and for the unit of measurement take the segment AB. Find the coordinates of all the vertices of this hexagon. How many solutions does this problem have?

8. In a plane the points $A(0, 0)$, $B(x_1, y_1)$, and $D(x_2, y_2)$ are given. What coordinates must the point C have so that the quadrangle $ABCD$ will be a parallelogram?

[1] For the unit of measurement of area we take the area of a square whose sides are equal to the unit of measurement on the axes.

5. Relations Connecting Coordinates

If both coordinates of a point are known, then its position in the plane is fully defined. What can one say about the position of a point if only one of its coordinates is known? For example, where do all the points whose abscissas are equal to three lie? Where are the points one of whose coordinates is equal to three located?

Generally speaking, specifying one of the two coordinates determines some curve. Indeed, the plot of Jules Verne's novel, *The Children of Captain Grant*, was based on this fact. The heroes of the book knew only one coordinate of the place where a shipwreck had occurred (the latitude), and therefore, in order to examine all possible locations, they were forced to circle the earth along an entire parallel — the line for each of whose points the latitude was equal to 37°11′.

A relation between coordinates usually defines not merely a point but a *set* (collection) of points. For example, if one marks off all points whose abscissas are equal to their ordinates, that is, those points whose coordinates satisfy the relation

$$x = y,$$

one gets a straight line: the bisector of the first and third coordinate angles (Fig. 8).

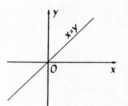

Fig. 8

Sometimes, instead of a "set of points" one speaks of a "locus of points." For example, the locus of points whose coordinates satisfy the equation

$$x = y$$

is, as we have said, the bisector of the first and third coordinate angles.

One should not suppose, however, that every relation between the coordinates necessarily gives a line in the plane. For example, you can easily see that the relation $x^2 + y^2 = 0$ defines a single point: the origin. The relation $x^2 + y^2 = -1$ is not satisfied

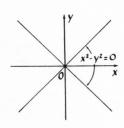

Fig. 9

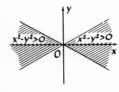

Fig. 10

by the coordinates of any point (it defines the so-called empty set).

The relation

$$x^2 - y^2 = 0$$

leads to a pair of mutually perpendicular straight lines (Fig. 9). The relation $x^2 - y^2 > 0$ gives a whole region (Fig. 10).

EXERCISES

1. Try to decide by yourself which sets of points are defined by these relations:

(a) $|x| = |y|$;

(b) $x/|x| = y/|y|$;

(c) $|x| + x = |y| + y$;

(d) $[x] = [y]$;[1]

(e) $x - [x] = y - [y]$;

(f) $x - [x] > y - [y]$. (The answer to Exercise 1f is given in a figure on page 73.)

2. A straight path separates a meadow from a field. A pedestrian travels along the path at a speed of 5 km/hr, through the meadow at a speed of 4 km/hr, and through the field at a speed of 3 km/hr. Initially, the pedestrian is on the path. Draw the region which the pedestrian can cover in 1 hour.

3. The plane is divided by the coordinate axes into four quadrants. In the first and third quadrants (including the coordinate axes) it is possible to travel at the speed a, and in the second and fourth (excluding the coordinate axes) one can travel at the speed b. Draw the set of points which can be reached from the origin over a given amount of time if:

(a) the speed a is twice as great as b;

(b) the speeds are connected by the relation

$$a = b\sqrt{2}.$$

---

[1] The symbol $[x]$ denotes the whole part of the number x, that is, the largest whole number not exceeding x. For example, $[3.5] = 3$, $[5] = 5$, $[-2.5] = -3$.

18

6. The Distance Between Two Points

We are now able to speak of points using numerical terminology. For example, we do not need to say: Take the point located three units to the right of the y-axis and five units beneath the x-axis. It suffices to say simply: Take the point $(3, -5)$.

We have already said that this creates definite advantages. Thus, we can transmit a figure consisting of points by telegraph or by a computer, which does not understand sketches but does understand numbers.

In the preceding section, with the aid of relations among numbers, we have given some sets of points in the plane. Now let us try to translate other geometrical concepts and facts into the language of numbers.

We shall begin with a simple and ordinary problem: to find the distance between two points in the plane.

As always, we shall suppose that the points are given by their coordinates; and thus our problem reduces to the following: to find a rule according to which we will be able to calculate the distance between two points if we know their coordinates. In the derivation of this rule, of course, resorting to a sketch will be permitted, but the rule itself must not contain any reference to the sketch but must show only how and in what order one should operate with the given numbers — the coordinates of the points — in order to obtain the desired number — the distance between the points.

Possibly, to some of our readers this approach to the solution of the problem will seem strange and forced. What could be simpler, they will say, for the points are given, even though by their coordinates. Draw the points, take a ruler and measure the distance between them.

This method sometimes is not so bad. But suppose again that you are dealing with a digital computer. There is no ruler in it, and it does not draw; but it is able to compute so quickly[1] that this causes it no

[1]A modern computing machine carries out tens of thousands of operations of addition and multiplication per second.

difficulty. Notice that our problem is so set up that the rule for calculating the distance between two points will consist of commands which the machine can carry out.

It is better first to solve the problem which we have posed for the special case where one of the given points lies at the origin of coordinates. Begin with some numerical examples: find the distance from the origin of the points $(12, 5)$, $(-3, 15)$, and $(-4, -7)$.

Hint. Use the Pythagorean theorem.

Now write down a general formula for calculating the distance of a point from the origin of coordinates.

Answer. The distance of the point $M(x, y)$ from the origin of coordinates is defined by the formula

$$d(O, M) = \sqrt{x^2 + y^2}.$$

Obviously, the rule expressed by this formula satisfies the conditions set above. In particular, it can be used to calculate with machines that can multiply numbers, add them, and extract roots.

Let us now solve the general problem.

Problem. Given two points in the plane, $A(x_1, y_1)$ and $B(x_2, y_2)$, find the distance $d(A, B)$ between them.

Solution. Let us denote by A_1, B_1, A_2, B_2 (Fig. 11) the projections of the points A and B on the coordinate axis.

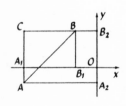

Fig. 11

Let us denote the point of intersection of the straight lines AA_1 and BB_2 by the letter C. From the right triangle ABC we get, from the Pythagorean theorem,[1]

$$d^2(A, B) = d^2(A, C) + d^2(B, C). \qquad (*)$$

But the length of the segment AC is equal to the length of the segment A_2B_2. The points A_2 and B_2

[1] By $d^2(A, B)$ we mean the square of the distance $d(A, B)$.

lie on the axis Oy and have the coordinates y_1 and y_2, respectively. According to the formula obtained on page 11, the distance between them is equal to $|y_1 - y_2|$.

By an analogous argument we find that the length of the segment BC is equal to $|x_1 - x_2|$. Substituting the values of AC and BC that we have found in the formula (*), we obtain:

$$d^2(A, B) = (x_1 - x_2)^2 + (y_1 - y_2)^2.$$

Thus, $d(A, B)$ — the distance between the points $A(x_1, y_1)$ and $B(x_2, y_2)$ — is computed by the formula

$$d(A, B) = \sqrt{(x_1 - x_2)^2 + (y_1 - y_2)^2}\,.$$

Let us note that our entire argument is valid not only for the disposition of points shown in Fig. 11 but for any other.

Make another sketch (for example, take the point A in the first quadrant and the point B in the second) and convince yourself that the entire argument can be repeated word for word without even changing the designations of the points.

Note also that the formula on page 10 for the distance between points on the straight line can be written in an analogous form:[1]

$$d(A, B) = \sqrt{(x_1 - x_2)^2}\,.$$

[1] We use the fact that

$$\sqrt{x^2} = |x|$$

(keep in mind the arithmetic value of the root). An inaccurate use of this rule (sometimes people mistakenly calculate that $\sqrt{x^2} = x$) can lead to an incorrect conclusion. As an example, we give a chain of reasoning containing such an inaccuracy and invite you to try to discover it:

$$1 - 3 = 4 - 6 \Rightarrow 1 - 3 + \tfrac{9}{4} = 4 - 6 + \tfrac{9}{4}$$
$$\Rightarrow (1 - \tfrac{3}{2})^2 = (2 - \tfrac{3}{2})^2 \Rightarrow \sqrt{(1 - \tfrac{3}{2})^2}$$
$$= \sqrt{(2 - \tfrac{3}{2})^2} \Rightarrow 1 - \tfrac{3}{2} = 2 - \tfrac{3}{2} \Rightarrow 1 = 2.$$

1. In the plane three points $A(3, -6)$, $B(-2, 4)$ and $C(1, -2)$ are given. Prove that these three points lie on the same line. (**Hint.** Show that one of the sides of the "triangle" ABC is equal to the sum of the other two sides.)

2. Apply the formula for the distance between two points to prove the well-known theorem: In a parallelogram the sum of the squares of the sides is equal to the sum of the squares of the diagonals. (**Hint.** Take one of the vertices of the parallelogram to be the origin of coordinates and use the result of Problem 3 on page 16. You will see that the proof of the theorem reduces to checking a simple algebraic identity. Which?)

3. Using the method of coordinates, prove the following theorem: if $ABCD$ is a rectangle, then for an arbitrary point M the equality

$$AM^2 + CM^2 = BM^2 + DM^2$$

is valid. What is the most convenient way of placing the coordinate axes?

7. Defining Figures

In Section 5 we introduced some examples of relations between the coordinates that define figures on the plane. We shall now go further into the study of geometrical figures using relations between numbers.

We view each figure as a collection of points, the points on the figure; and to give a figure will mean to establish a method of telling whether or not a point belongs to the figure under consideration.

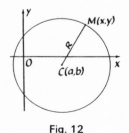

Fig. 12

In order to find such a method — for example, for the circle — we use the definition of the circle as the set of points whose distance from some point C (the center of the circle) is equal to a number R (the radius). This means that in order for the point $M(x, y)$ (Fig. 12) to lie on the circle with the center

$C(a, b)$, it is necessary and sufficient that $d(M, C)$ be equal to R.

Let us recall that the distance between points is defined by the formula

$$d(A, B) = \sqrt{(x_1 - x_2)^2 + (y_1 - y_2)^2}.$$

Consequently, the condition that the point $M(x, y)$ lie on the circle with center $C(a, b)$ and radius R is expressed by the relation $\sqrt{(x - a)^2 + (y - b)^2} = R$, which can be rewritten in the form

$$(x - a)^2 + (y - b)^2 = R^2. \qquad \text{(A)}$$

Thus, in order to check whether or not a point lies on a circle, we need merely check whether or not the relation (A) is satisfied for this point. For this we must substitute the coordinates of the given points for x and y in (A). If we obtain an equality, then the point lies on the circle; otherwise, the point does not lie on the circle. Thus, knowing equation (A), we can determine whether or not a given point in the plane lies on the circle. Therefore equation (A) is called the *equation of the circle* with center $C(a, b)$ and radius R.

EXERCISES

1. Write the equation of the circle with center $C(-2, 3)$ and radius 5. Does this circle pass through the point $(2, -1)$?

2. Show that the equation

$$x^2 + 2x + y^2 = 0$$

specifies a circle in the plane. Find its center and radius. (**Hint.** Put the equation in the form

$$(x^2 + 2x + 1) + y^2 = 1,$$

or

$$(x + 1)^2 + y^2 = 1.)$$

3. What set of points is specified by the equation $x^2 + y^2 \leq 4x + 4y$?

(**Solution.** Rewrite the inequality in the form

$$x^2 - 4x + 4 + y^2 - 4y + 4 \leq 8,$$

or

$$(x - 2)^2 + (y - 2)^2 \leq 8.)$$

As is now clear, this relation says that the distance of any point in the desired set to the point $(2, 2)$ is less than or equal to $\sqrt{8}$. It is evident that the points satisfying this condition fill the circle with radius $\sqrt{8}$ and center at $(2, 2)$. Since equality is permitted in the relation, the boundary of the circle also belongs to the set.)

We have seen that a circle in the plane can be given by means of an equation. In the same way one can specify other curves; but their equations, of course, will be different.

We have already said that the equation $x^2 - y^2 = 0$ specifies a pair of straight lines (see page 16). Let us examine this somewhat more closely. If $x^2 - y^2 = 0$, then $x^2 = y^2$ and consequently, $|x| = |y|$. On the other hand, if $|x| = |y|$, then $x^2 - y^2 = 0$; therefore, these relations are equivalent. But the absolute value of the abscissa of a point is the distance of the point from the axis Oy, and the absolute value of the ordinate is its distance from the axis Ox. This means that the points for which $|x| = |y|$ are equidistant from the coordinate axes, that is, lie on the two bisectors of the angles formed by these axes. It is clear, conversely, that the coordinates of an arbitrary point on each of these two bisectors satisfy the relation $x^2 = y^2$. We shall say, therefore, that the equation of the points on these two bisectors is the equation $x^2 - y^2 = 0$.

You know, of course, other examples of curves that are given by means of an equation. For example, the equation $y = x^2$ is satisfied by all the points of a parabola with vertex at the origin, and only by these points. The equation $y = x^2$ is called the *equation* of this *parabola*.

In general, by the *equation* of a *curve* we mean that equation which becomes an identity whenever the coordinates of any point on the curve are substituted for x and y in the equation, and which is not satisfied if one substitutes the coordinates of a point not lying on the curve.

For example, without even knowing what the curve specified by the equation

$$(x^2 + y^2 + y)^2 = x^2 + y^2 \qquad (*)$$

looks like, we can say that it passes through the origin, since the numbers (0, 0) satisfy the equation. However, the point (1, 1) does not lie on this curve, since $(1^2 + 1^2 + 1^2)^2 \neq 1^2 \neq 1^2$.

If you are interested in seeing what the curve specified by this equation looks like, look at Fig. 13. This curve is called a *cardioid* since it has the shape of a heart.

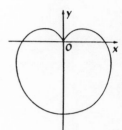

Fig. 13

If a computer could feel affection toward someone, it would probably transmit the figure of a heart in the form of an equation to him; but on the other hand, perhaps it would give a mathematical "bouquet" — the equation of the curves shown in Fig. 14. As you see, these curves are really quite similar to flowers. We shall write out the equations of these mathematical flowers when you have become acquainted with polar coordinates.

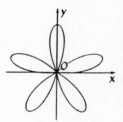

Fig. 14*a*

8. We Begin to Solve Problems

The translation of geometrical concepts into the language of coordinates permits us to consider algebraic problems in place of geometric ones. It turns out that after such a translation the majority of problems connected with lines and circles lead to equations of the first and second degree; and there are simple general formulas for the solution of these equations. (It should be noted that in the seventeenth century, when the method of coordinates was devised,

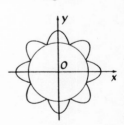

Fig. 14*b*

25

the art of solving algebraic equations has reached a very high level. By this time, for example, mathematicians had learned how to solve arbitrary equations of the third and fourth degree. The French philosopher René Descartes, in disclosing the method of coordinates was able to boast: "I have solved all problems" — meaning the geometric problems of his time.)

We shall now illustrate by a simple example the reduction of geometric problems to algebraic ones.

Problem. Given the triangle ABC; find the center of the circle circumscribed about this triangle.

Solution. Let us take the point A as the origin and direct the x-axis from A to B. Then the point B will have the coordinates $(c, 0)$, where c is the length of the segment AB. Let the point C have the coordinates (q, h), and let the center of the desired circle have the coordinates (a, b). The radius of this circle we denote by R. We write down in coordinate language that the points $A(0, 0)$, $B(c, 0)$, and $C(q, h)$ lie on the desired circle:

$$a^2 + b^2 = R^2,$$

$$(c - a)^2 + b^2 = R^2,$$

$$(q - a)^2 + (h - b)^2 = R^2.$$

These conditions express the fact that the distance of each of the points $A(0, 0)$, $B(c, 0)$, and $C(q, h)$ from the center of the circle (a, b) is equal to the radius. One also obtains these conditions easily if one writes down the equations of the unknown circle (the circle with its center at (a, b) and radius R), that is,

$$(x - a)^2 + (y - b)^2 = R^2,$$

and then substitutes the coordinates of the points A, B, and C, lying on this circle, for x and y.

This system of three equations with three unknowns is easily solved, and we get

$$a = \frac{c}{2}, \qquad b = \frac{q^2 + h^2 - cq}{2h}$$

$$R = \frac{\sqrt{(q^2 + h^2)[(q - c)^2 + h^2]}}{2h}.$$

This problem is solved, since we have found the coordinates of the center.[1]

Let us note that at the same time we have obtained a formula for calculating the radius of the circle circumscribed about a triangle. We can simplify this formula if we note that $\sqrt{q^2 + h^2} = d(A, C)$, $\sqrt{(q - c)^2 + h^2} = d(B, C)$, and the dimension h is equal to the altitude of the triangle ABC dropped from the vertex C. If we denote the lengths of the sides BC and AC of the triangle by a and b, respectively, then the formula for the radius assumes the beautiful and useful form:

$$R = \frac{ab}{2h}.$$

One can remark further that $hc = 2S$, where S is the area of the triangle ABC; and thus we can write our formula in the form:

$$R = \frac{abc}{4S}.$$

Now we wish to show you a problem which is interesting because its geometric solution is quite complicated, but if we translate it into the language of coordinates, its solution becomes quite simple.

Problem. Given two points A and B in the plane, find the locus of points M whose distance from A is twice as great as from B.

[1]Notice that in the solution of this problem we have not resorted to a sketch.

Solution. Let us choose a system of coordinates on the plane such that the origin is located at the point A and the positive part of the x-axis lies along AB. We take the length of AB as the unit of length. Then the point A will have coordinates $(0, 0)$, and the point B will have the coordinates $(1, 0)$. The coordinates of the point M we denote by (x, y). The condition $d(A, M) = 2d(B, M)$ is written in coordinates as follows:

$$\sqrt{x^2 + y^2} = 2\sqrt{(x - 1)^2 + y^2}.$$

We have obtained the equation of the desired locus of points. In order to determine what this locus looks like, we transform the equation into a more familiar form. Squaring both sides, removing the parentheses, and transposing like terms, we get the equation

$$3x^2 - 8x + 4 + 3y^2 = 0.$$

This equation can be rewritten as follows:

$$x^2 - \tfrac{8}{3}x + \tfrac{16}{9} + y^2 = \tfrac{4}{9},$$

or

$$(x - \tfrac{4}{3})^2 + y^2 = (\tfrac{2}{3})^2.$$

You already know that this equation is the equation of the circle with center at the point $(\tfrac{4}{3}, 0)$ and radius equal to $\tfrac{2}{3}$. This means that our locus of points is a circle.

For our solution it is inessential that $d(A, M)$ be specifically two times as large as $d(B, M)$, since in fact we have solved a more general problem: We have proved that *the locus of points M, the ratio of whose distances to the given points A and B is constant:*

$$\frac{d(A, M)}{d(B, M)} = k \qquad \qquad \text{(*)}$$

(k is a given positive number not equal to 1), *is a circle.*[1]

[1] We have excluded the case $k = 1$; you of course know that in this case the locus (*) is a straight line (the point M is equidistant from A and B). Prove this analytically.

In order to convince yourself of the power of the method of coordinates, try to solve this same problem geometrically. (**Hint.** Draw the bisectors of the internal and external angles of the triangle AMB at the point M. Let K and L be the points of intersection of these bisectors with the line AB. Show that the position of these points does not depend on the choice of the point M in the desired locus of points. Show that the angle KML is equal to 90°.)

We should remark that even the ancient Greeks knew how to cope with such problems. The geometrical solution of this problem is found in the treatise "On Circles" by the ancient Greek mathematician Apollonius (second century B.C.).

Solve the following problem by yourself:

Find the locus of points M the difference of the squares of whose distances from two given points A and B is equal to a given value c. For what values of c does the problem have a solution?

9. Other Systems of Coordinates

In the plane, coordinate systems other than a rectangular Cartesian one are often used. In Fig. 15 an oblique Cartesian system of coordinates is depicted. It is clear from the picture how the coordinates of a point are defined in such a system. In some cases it is necessary to take different units of measurement along the coordinate axes.

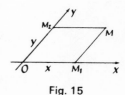

Fig. 15

There are coordinates that are more essentially different from rectangular Cartesian ones. An example of these coordinates is the system of polar coordinates to which we have already referred.

The polar coordinates of a point in the plane are defined in the following way.

A number axis is chosen in the plane (Fig. 16). The origin of coordinates of this axis (the point O) is called the *pole*, and the axis itself is the *polar axis*. To define the position of a point M it suffices to indicate two numbers — ρ, the *polar radius* (the distance of the point from the pole), and ϕ, the *polar angle*[1] (the angle of rotation from the polar axis to the half-line OM). In our sketch the polar

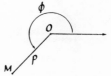

Fig. 16

[1] ρ and ϕ are the Greek letters *rho* and *phi*.

radius is equal to 3.5 and the polar angle ϕ is equal to 225° or $5\pi/4$.[1]

Thus, in a polar system of coordinates the position of a point is specified by two numbers, which indicate the direction in which the point is to be found and the distance to this point. Such a method of defining position is quite simple and is frequently used. For example, in order to explain the way to someone who is lost in a forest, one might say: "Turn east (the direction) at the burnt pine (the pole), go two kilometers (the polar distance) and there you will find the lodge (the point)."

Anyone who has traveled as a tourist will easily see that going along an azimuth is based on the same principle as polar coordinates.

Polar coordinates, like Cartesian ones, can be used to specify various sets of points in the plane. The equation of a circle in polar coordinates, for example, turns out to be quite simple if the center of the circle is taken as the pole. If the radius of the circle is equal to R, then the polar radius of any point of the circle (and only of points on this circle) is also equal to R, so that the equation of this circle has the form

$$\rho = R,$$

where R is some constant quantity.

What set of points is determined by the equation $\phi = \alpha$, where α is some constant number (for example, $\frac{1}{2}$ or $3\pi/2$)? The answer is clear: the points for which ϕ is constant and equal to α are the points on the half-line directed outward from the pole at an angle α to the polar axis. For example, if $\alpha = \frac{1}{2}$, this half-line passes along at an angle equal approximately to 28°,[2] and if $\alpha = 3\pi/2$,

[1] For measuring angles in the polar system of coordinates we use either the degree or the *radian* — the central angle formed by an arc with length 1 of a circle of radius 1. A full angle of 360° formed by an entire circle (of radius 1) acquires the radian measure 2π, a 180° angle — the measure π, a right angle — the measure $\pi/2$, a 45° angle — the measure $\pi/4$, and so on. A radian is equal to $180°/\pi \approx 180°/3.14 \approx 57°17'45''$. It turns out that in many problems radian measure is significantly more convenient than degree measure.

[2] Let us recall that the number serving as the coordinate ϕ must be interpreted as the radian measure of the angle (see the previous note). An angle of $\frac{1}{2}$ radian is approximately equal to 28°; an angle of $3\pi/2$ radians is equal to 270° (exactly).

the half-line is directed vertically downward; that is, the angle between the positive direction of the axis and the half-line is equal to 270°.

Let us take two more examples. The equation

$$\rho = \phi$$

describes a spiral (Fig. 17a). In fact, for $\phi = 0$ we have $\rho = 0$ (the pole); and as ϕ grows, the quantity ρ also grows, so that a point traveling around the pole (in a counterclockwise direction) simultaneously gets farther away from it.

Another spiral is described by the equation

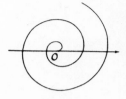

Fig. 17a

$$\rho = \frac{1}{\phi}$$

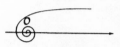

Fig. 17b

(Fig. 17b). In this case, for ϕ close to 0 the value of ρ is large; but with the growth of ϕ the value of ρ diminishes and is small for large ϕ. Therefore, the spiral winds into the point 0 as ϕ grows large without bound.

The equations of curves in the polar system of coordinates might be more difficult for you to understand, particularly if you have not studied trigonometry. If you are somewhat familiar with this subject, try to figure out what sets of points are determined by these relations:

$$\rho = \sin \phi, \qquad \rho(\cos \phi + \sin \phi) + 1 = 0.$$

The polar system of coordinates is in some cases more convenient than the Cartesian. Here, for example, is the equation of the cardioid in polar coordinates (see Section 7):

$$\rho = 1 - \sin \phi.$$

Some knowledge of trigonometry will enable you to visualize the curve somewhat more easily from this equation than from the equation of the curve in Cartesian coordinates. Using polar coordinates, you will also be able to describe the flowers shown in Fig. 14 by the following equations, which are quite simple:

$$\rho = \sin 5\phi \qquad \text{(Fig. 14a)},$$
$$(\rho - 2)(\rho - 2 - |\cos 3\phi|) = 0 \qquad \text{(Fig. 14b)}.$$

31

We have not spoken about one-to-one correspondences between the points on the plane and polar coordinates. This is because such a one-to-one correspondence simply does not exist. In fact, if you add an arbitrary integral multiple of 2π (that is, of the angle 360°) to the angle ϕ, then the direction of a half-line at the angle ϕ to the polar axis is clearly not changed. In other words, points with the polar coordinates ρ, ϕ and ρ, $\phi + 2k\pi$, where $\rho > 0$ and k is any integer, coincide. We wish to introduce still another example where the correspondence is not one-to-one.

In the Introduction we remarked that it is possible to define coordinates on curves, and in Chapter 1 we examined coordinates on the simplest kind of curve: a straight line. Now we shall show that it is possible to devise coordinates for still another curve: a circle. For this, as in Chapter 1, we choose some point on the circle as the origin (the point O in Fig. 18). As usual, we shall take clockwise motion as the positive direction of motion on the circle. The unit of measure on the circle can likewise be chosen in a natural manner: We take the radius of the circle as the unit of measure. Then the coordinate of the point M on the circle will be the length of the arc OM, taken with a plus sign if the rotation from O to M is in the positive direction and with a minus sign in the opposite case.

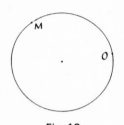

Fig. 18

Immediately an important difference becomes apparent between these coordinates and the coordinates of a point on the line: here there is no one-to-one correspondence between numbers (coordinates) and points. It is clear that for each number there is defined exactly one point on the circle. However, suppose the number a is given; in order to find the point on the circle corresponding to it (that is, the point with the coordinate a), one must lay off on the circle an arc of length a radii in the positive direction if the number a is positive and in the negative direction if a is negative. Thus, for example, the point with coordinate 2π coincides with the origin. In our example the point O is obtained when the coordinate is equal to zero and when it is equal to 2π. Thus in the other direction the correspondence is not single-valued; that is, to the same point there corresponds more than one number. One easily sees that to each point on the circle there corresponds an infinite set of numbers. [1]

[1] Note that the coordinate introduced for points on the circle coincides with the angle ϕ of the polar system of coordinates, if the latter is measured in radians. Thus, the failure of polar coordinates to be one-to-one is once again illustrated here.

CHAPTER 3

The Coordinates of a Point in Space

10. Coordinate Axes and Planes

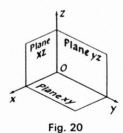

For the definition of the position of a point in space it is necessary to take not two number axes (as in the case of the plane) but three: the x-axis, the y-axis, and the z-axis. These axes pass through a common point — the *origin of coordinates* O — in such a manner that any two of them are mutually perpendicular. The direction of the axes is usually chosen so that the positive half of the x-axis will coincide with the positive half of the y-axis after a 90° rotation counterclockwise if one is looking from the positive part of the z-axis (Fig. 19).

Fig. 19

In space, it is convenient to consider, in addition to the coordinate axes, the *coordinate planes*, that is, the planes passing through any two coordinate axes. There are three such planes (Fig. 20).

The xy-plane (passing through the x- and y-axes) is the set of points of the form $(x, y, 0)$, where x and y are arbitrary numbers.

The xz-plane (passing through the x- and z-axes) is the set of points of the form $(x, 0, z)$, where x and z are arbitrary numbers.

Fig. 20

34

The yz-plane (passing through the y- and z-axes) is the set of points of the form $(0, y, z)$, where y and z are arbitrary numbers.

Now for each point M in space one can find three numbers x, y, and z that will serve as its coordinates.

In order to find the first number x, we construct through the point M the plane parallel to the coordinate plane yz (the plane so constructed will also be perpendicular to the x-axis). The point of intersection of this plane with the x-axis (the point M_1 in Figure 21a) has the coordinate x on this axis. This number x is the coordinate of the point M_1 on the x-axis and is called the x-coordinate of the point M.

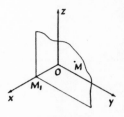

Fig. 21a

In order to find the second coordinate, we construct through the point M the plane parallel to the xz-plane (perpendicular to the y-axis), and find the point M_2 on the y-axis (Fig. 21b). The number y is the coordinate of the point M_2 on the y-axis and is called the y-coordinate of the point M.

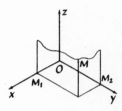

Fig. 21b

Analogously, by constructing through the point M the plane parallel to the xy-plane (perpendicular to the z-axis), we find the number z — the coordinate of the point M_3 (Fig. 21c) on the z-axis. This number z is called the z-coordinate of the point M.

In this way, we have defined for each point in space a triple of numbers to serve as coordinates: the x-coordinate, the y-coordinate, and the z-coordinate.

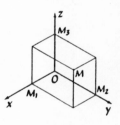

Fig. 21c

Conversely, to each triple of numbers (x, y, z) in a definite order (first x, then y, then z) one can place in correspondence a definite point M in space. For this one must use the construction already described, carrying it out in the reverse order: mark off on the axes the points M_1, M_2, and M_3 having the coordinates x, y, and z, respectively, on these axes, and then construct through these points the planes parallel to the coordinate planes. The point of intersection of these three planes will be the desired point M. It is evident that the numbers (x, y, z) will be its coordinates.

In this way, we have set up a one-to-one correspondence[1]

[1] For the definition of a one-to-one correspondence see page 6.

between the points of space and ordered triples of numbers (the coordinates of these points).

Mastering coordinates in space will be more difficult for you than mastering coordinates on a plane was: for the study of coordinates in space requires some knowledge of solid geometry. The material necessary for understanding space coordinates, which you will understand easily on account of their simplicity and obviousness, is given a somewhat more rigorous foundation in courses in solid geometry.

In such a course one shows that the points M_1, M_2, and M_3, constructed as the points of intersection of the coordinate axes with the planes drawn through the point M parallel to the coordinate planes, are the projections of the point M on the coordinate axes, that is, that they are the bases of the perpendiculars dropped from the point M to the coordinate axes. Thus for coordinates in space we can give a definition analogous to the definition of coordinates of a point in the plane:

The *coordinates* of a point M in space are the coordinates on the coordinate axes of the projections of the point M onto these axes.

One can show that many formulas derived for the plane become valid for space with only a slight change in their form. Thus, for example, the distance between two points $A(x_1, y_1, z_1)$ and $B(x_2, y_2, z_2)$ can be calculated by the formula

$$d(A, B) = \sqrt{(x_1 - x_2)^2 + (y_1 - y_2)^2 + (z_1 - z_2)^2}.$$

(The derivation of this formula is quite similar to the derivation of the analogous formula for the plane. Try to carry it out by yourself.)

In particular, the distance between a point $A(x, y, z)$ and the origin is expressed by the formula

$$d(O, A) = \sqrt{x^2 + y^2 + z^2}.$$

EXERCISES

1. Take these eight points: $(1, 1, 1)$, $(1, 1, -1)$, $(1, -1, 1)$, $(1, -1, -1)$, $(-1, 1, 1)$, $(-1, 1, -1)$, $(-1, -1, 1)$, $(-1, -1, -1)$. Which of these points is farthest from the point $(1, 1, 1)$? Find the distance of this point from $(1, 1, 1)$. Which points lie closest to the point $(1, 1, 1)$? What is the distance of these points from $(1, 1, 1)$?

2. Draw a cube. Direct the coordinate axes along the three edges adjacent to any one vertex. Take the edge of the cube as the unit of measurement. Denote the vertices of the cube by the letters A, B, C, D, A_1, B_1, C_1, D_1, as in Fig. 22.

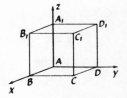

Fig. 22

(a) Find the coordinates of the vertices of the cube.

(b) Find the coordinates of the midpoint of the edge CC_1.

(c) Find the coordinate of the point of intersection of the diagonals of the face AA_1B_1B.

3. What is the distance from the vertex $(0, 0, 0)$ of the cube in Problem 1 to the point of intersection of the diagonals of the edge BB_1C_1C?

4. Which of the following points

$$A(1, 0, 5), \quad B(3, 0, 1), \quad C(\tfrac{1}{3}, \tfrac{3}{4}, \tfrac{2}{5}),$$
$$D(\tfrac{7}{5}, \tfrac{1}{2}, \tfrac{3}{2}), \quad E(\tfrac{2}{5}, -\tfrac{1}{2}, 0), \quad F(1, \tfrac{1}{2}, \tfrac{1}{3})$$

do you think lie inside the cube in Problem 1, and which lie outside?

5. Write down the relations which the coordinates of the points lying inside and on the boundary of the cube in Problem 1 satisfy.

(**Answer.** The coordinates x, y, and z of the points lying inside our cube and on its boundary can take on all numerical values from zero to one inclusive; that is, they satisfy the relations

$$0 \leq x \leq 1,$$
$$0 \leq y \leq 1,$$
$$0 \leq z \leq 1.)$$

11. Defining Figures in Space

Just as in the plane, coordinates in space enable us to define by means of numbers and numerical relations not only points but also sets of points such as curves and surfaces. We can, for example, define the set of points by specifying two coordinates — say the x-coordinate and the y-coordinate — and taking the third one arbitrarily. The conditions $x = a$, $y = b$, where a and b are given numbers (for example, $a = 5$, $b = 4$), define in space a straight line parallel to the z-axis (Fig. 23). All of the points of this line have the same x-coordinate and y-coordinate; their z-coordinates assume arbitrary values.

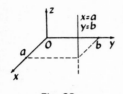

Fig. 23

In exactly the same way the conditions

$$y = b, \qquad z = c$$

define a straight line parallel to the x-axis; and the conditions

$$z = c, \qquad x = a$$

define a straight line parallel to the y-axis.

Here is an interesting question: What set of points is obtained if one specifies only one coordinate, for example,

$$z = 1?$$

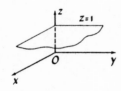

Fig. 24

The answer is clear from Fig. 24: it is the plane parallel to the xy-coordinate plane (that is, the plane passing through the x- and y-axes) and at a distance of 1 from it in the direction of the positive semiaxis z.

Let us take some more examples showing how one can define various sets of points in space with the aid of equations and other relations between the coordinates.

1. Let us examine the equation

$$x^2 + y^2 + z^2 = R^2. \tag{*}$$

As the distance of the point (x, y, z) from the origin of coordinates is given by the expression $\sqrt{x^2 + y^2 + z^2}$, it is clear that, translated into geometrical language, the relation (*) indicates that the point with the co-ordinates (x, y, z) satisfying this relation is located at a distance R from the origin of coordinates. This means that the set of all points for which the relation (*) is satisfied is the surface of a sphere — the sphere with center at the origin and with radius R.

2. Where are the points located whose coordinates satisfy the relation

$$x^2 + y^2 + z^2 < 1?$$

Answer. Since this relation means that the distance of the point (x, y, z) from the origin is less than 1, the desired set is the set of points lying within the sphere with center at the origin and with radius equal to 1.

3. What set of points is specified by the following equation?

$$x^2 + y^2 = 1. \qquad (**)$$

Let us examine first only the points on the xy-plane satisfying this relation, that is, the points for which $z = 0$. Then this equation, as we have seen before (page 21), defines a circle with center at the origin and radius equal to 1. Each of these points has its z-coordinate equal to 0, and the x- and y-coordinates satisfy the relation (**). For example, the point $(\frac{3}{5}, \frac{4}{5}, 0)$ satisfies this equation (**) (Fig. 25). More-over, knowing this one point, we can immediately find many other points satisfying the same equation. In fact, since z is not present in the equation (**), the point $(\frac{3}{5}, \frac{4}{5}, 10)$, the point $(\frac{3}{5}, \frac{4}{5}, -5)$, and in general the points $(\frac{3}{5}, \frac{4}{5}, z)$, where the value of the z-coordinate is absolutely arbitrary, satisfy the equation. All of these points lie on the straight line passing through the point $(\frac{3}{5}, \frac{4}{5}, 0)$ parallel to the z-axis.

In this way each point $(x^*, y^*, 0)$ of our circle in the xy-plane gives rise to many points satisfying equation (**) — the points on the straight line passing

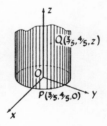

Fig. 25

through this point parallel to the z-axis. All of the points of this line will have the same x- and y-coordinates as the point on the circle, but z can be an arbitrary number, that is, they will be points of the form (x^*, y^*, z). But since z does not enter into equation (**) and the numbers $(x^*, y^*, 0)$ satisfy the equation, the numbers (x^*, y^*, z) also satisfy equation (**) for any z. It is clear that in this way one obtains every point satisfying equation (**).

Thus, the set of points determined by equation (**) is obtained in the following manner: Take the circle with its center at the origin and radius 1 lying in the xy-plane, and through each point of this circle construct a straight line parallel to the z-axis. We thus obtain a *cylindrical surface* (Fig. 25).

4. We have seen that a single equation generally defines a surface in space. But this is not always so. For example, the equation $x^2 + y^2 = 0$ is satisfied only by the points of a line — the z-axis — since it follows from the equation that x and y are equal to zero, and all points for which these coordinates are equal to zero lie on the z-axis. The equation $x^2 + y^2 + z^2 = 0$ describes a single point (the origin); but the equation $x^2 + y^2 + z^2 = -1$ is satisfied by no points at all, and so it corresponds to the empty set.

5. What happens if we consider points whose coordinates satisfy not a single equation but a system of equations?

Let us examine such a system of questions:

$$\left.\begin{array}{r} x^2 + y^2 + z^2 = 4, \\ z = 1. \end{array}\right\} \qquad (***)$$

The points satisfying the first equation fill up the surface of a sphere of radius 2 and center at the origin. The points satisfying the second equation fill up the plane parallel to the xy-plane and located at a distance of 1 from it on the positive side of the z-axis. The points satisfying both the first and second equation must therefore lie both on the sphere $x^2 + y^2 + z^2 = 4$ and on the plane $z = 1$; that is, they lie on the curve of intersection. Thus, this system defines a circle: the curve of intersection of a sphere and a plane (Fig. 26).

We see that each of the equations of the system defines a surface, but both equations taken together define a line.

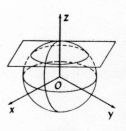

Fig. 26

Question. Which of the following points lie on the first surface, which on the second, and which on their line of intersection?

$$A(\sqrt{2}, \sqrt{2}, 0), \qquad B(\sqrt{2}, \sqrt{2}, 1),$$
$$C(\sqrt{2}, \sqrt{2}, \sqrt{2}), \qquad D(1, \sqrt{3}, 0),$$
$$E(0, \sqrt{3}, 1), \qquad F(-1, -\sqrt{2}, 1).$$

6. How can one give in space a circle located in the xz-plane with center at the origin and radius 1?

As you have already seen, the equation $x^2 + z^2 = 1$ defines a cylindrical surface in space. In order to get only the points on the circle we need, we must add to this equation the condition $y = 0$, distinguishing the points of the cylinder lying on the xz-plane from the rest of the points of the cylinder (Fig. 27). We therefore obtain the system

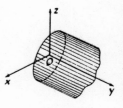

Fig. 27

$$\begin{cases} x^2 + z^2 = 1, \\ y = 0. \end{cases}$$

EXERCISES

1. What sets of points are defined in space by the relations: (a) $z^2 = 1$; (b) $y^2 + z^2 = 1$; (c) $x^2 + y^2 + z^2 = 1$?

2. Consider the three systems of equations:

(a) $\begin{cases} x^2 + y^2 + z^2 = 1, \\ y^2 + z^2 = 1; \end{cases}$

(b) $\begin{cases} x^2 + y^2 + z^2 = 1, \\ x = 0; \end{cases}$

(c) $\begin{cases} y^2 + z^2 = 1, \\ x = 0. \end{cases}$

Which of these define the same curve, and which define different ones?

3. How can one define in space the bisector of the angle $x0y$? What set of points in space will be given by the single equation $x = y$?

PART II

CHAPTER 1

Introduction

You now know something about the method of coordinates, and we can discuss some interesting things more closely related to modern mathematics.

1. Some General Considerations

Algebra and geometry, which most students today consider completely different subjects, are in fact quite closely related. With the aid of the method of coordinates it would be possible to present the entire school course in geometry without using a single sketch, using only numbers and algebraic operations. A course in plane geometry would begin with the words: "Let us define a point to be a pair of numbers (x, y). . . ." It would be further possible to define a circle as the set of points satisfying an equation of the form $(x - a)^2 + (y - b)^2 = R^2$. A straight line would be defined as the set of points satisfying an equation $ax + by + c = 0$, and so on. All geometric theorems would be converted in this approach into some algebraic relations.

Establishing a connection between algebra and geometry was, in essence, a revolution in mathematics. It restored mathematics as a single science, in which there is no "Chinese wall" between its individual parts. The French philosopher and mathematician René Descartes (1596–1650) is considered the creator of the method of coordinates. In the last part of his great philosophical treatise, published in 1637, a description of the method of coordinates was given, together with its application to the solution of geometric problems. The development of Descartes' idea led to the origin of a special branch of mathematics, now called analytic geometry.

The name itself indicates the fundamental idea of the theory. Analytic geometry is that branch of mathematics which solves geometric problems by analytical (that is, algebraic) means. Although analytic geometry today is a fully developed and perfected branch of mathematics, the idea on which it is based has given rise to new branches. One of these that has appeared and is being developed is algebraic geometry, in which the properties of curves and surfaces given by algebraic equations are studied. This field of mathematics can in no way be considered to be fully perfected. In fact, in recent years new fundamental results have been obtained in this field, and these have had a great influence upon other fields of mathematics as well.

2. Geometry as an Aid in Calculation

One aspect of the method of coordinates is of great importance in the solution of geometric problems: the analytic interpretation of geometric concepts and the translation of geometric forms and relations into the language of numbers. The other aspect of the method of coordinates, however — the geometric interpretation of numbers and of numerical relations — has acquired an equal significance. The distinguished mathematician Hermann Minkowski (1864–1909) used a geometric approach for the solution of equations in integers, and the mathematicians of his time were struck by how simple and clear some hitherto difficult questions in the theory of numbers turned out to be.

Here we shall take one quite simple example showing how geometry can help us to solve algebraic problems.

Problem. Let us consider the inequality

$$x^2 + y^2 \leq n,$$

where n is some integer. We would like to know how many solutions in integers this inequality has.

For small values of n, this question is easy to answer. For example, for $n = 0$, there is only one solution: $x = 0$, $y = 0$. For $n = 1$, there are four additional solutions: $x = 0$, $y = 1$; $x = 1$, $y = 0$; $x = 0$, $y = -1$; and $x = -1$, $y = 0$. Thus for $n = 1$, there will be five solutions in all.

For $n = 2$, there will be four more solutions besides the ones already enumerated: $x = 1$, $y = 1$; $x = -1, y = 1; x = 1, y = -1; x = -1, y = -1$. For $n = 2$, there are thus 9 solutions in all. Proceeding in this way, we can set up a table.

The Number n	The Number of Solutions N	The Ratio N/n
0	1	—
1	5	5
2	9	4.5
3	9	3
4	13	3.25
5	21	4.2
10	37	3.7
20	69	3.45
50	161	3.22
100	317	3.17

We see that the number of solutions N grows as n increases, but to guess the exact law for the change of N is quite difficult. One might conjecture in looking at the right column of the table that the ratio N/n converges to some number as n increases.

With the aid of a geometric interpretation we shall now show that this is in fact what occurs and that the ratio N/n converges to a number $\pi = 3.14159265....$

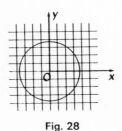

Fig. 28

Let us consider the pair of numbers (x, y) as a point on the plane (with abscissa x and ordinate y). The inequality $x^2 + y^2 \leq n$ means that the point (x, y) lies inside the circle K_n with radius $\sqrt{n}$ and with its center at the origin (Fig. 28). In this way, we see that our inequality has the same number of solutions in integers as there are points with integral coordinates lying inside the circle K_n.

It is geometrically clear that the points with integral coordinates are "uniformly distributed in the plane" and that to a unit square there will correspond one and only one such point. Therefore it is clear that the number of solutions must be approximately equal to the area of the circle. Thus we get the approximate formula:

$$N \approx \pi n.$$

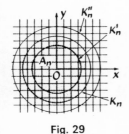

Fig. 29

We give a short proof of this formula. We first divide the plane into unit squares by straight lines parallel to the coordinate axes, letting the integral points be the vertices of these squares. Let there be N integral points inside the circle K_n. Let us place in correspondence with each of these points the unit square of which it is the upper right-hand vertex. The figure formed by these squares we denote by A_n (Fig. 29, the darkened part). It is evident that the area of A_n is equal to N (that is, to the number of squares in this figure).

Let us compare the area of this figure with the area of the circle K_n. Let us consider, together with the circle K_n, two other circles with the origin as center: the circle K_n' of radius $\sqrt{n} - \sqrt{2}$ and the circle K_n'' of radius $\sqrt{n} + \sqrt{2}$. The figure A_n lies entirely within the circle K_n'' and contains the circle K_n' entirely within itself. (Prove

48

this on your own, using the theorem that in a triangle the length of any side is less than the sum of the lengths of the other two sides.) Thus the area of A_n is greater than the area of K'_n and less than that of K''_n; that is,

$$\pi(\sqrt{n} - \sqrt{2})^2 < N < \pi(\sqrt{n} + \sqrt{2})^2.$$

From this we get our approximate formula $N \approx \pi n$ together with an estimate of its error:

$$|N - \pi n| < 2\pi(\sqrt{2n} + 1).$$

Let us now set up the analogous problem for three unknowns: How many solutions in integers does the following inequality have?

$$x^2 + y^2 + z^2 \leq n.$$

The answer is obtained quite easily if one again uses a geometric interpretation. The number of solutions to the problem is approximately equal to the volume of a sphere of radius $\sqrt{n}$ — that is, $\frac{4}{3}\pi n\sqrt{n}$. To obtain this result purely algebraically would be quite difficult.

3. The Need for Introducing Four-Dimensional Space

But what would happen if we had to find the number of integral solutions of the inequality

$$x^2 + y^2 + z^2 + u^2 \leq n,$$

in which there are four unknowns? In the solution of this problem for two and three unknowns, we have used a geometric interpretation. We have regarded a solution of the inequality for two unknowns — that is, a pair of numbers — as a point in the plane; we have regarded a solution for three unknowns — that is, a triple of numbers — as a point in space. Let us try to extend this method. Then the quadruple of numbers (x, y, z, u) must be considered as a point in some space having four dimensions (*four-dimensional space*). The inequality $x^2 + y^2 + z^2 + u^2 \leq n$ could then be viewed as the condition that the point (x, y, z, u) lie within the four-dimensional sphere with radius $\sqrt{n}$ and with its center at the origin. In addition, it would be necessary to decompose four-dimensional space into four-dimensional cubes. Finally, we would have to calculate the volume of the four-dimensional sphere.[1] In other words, we would have to begin to develop the geometry of four-dimensional space.

We shall not carry all of this out in this booklet. We shall be able to discuss only a very little bit of the subject here. As an introduction to four-dimensional space we shall discuss only the simplest figure in it: the four-dimensional cube.

Your interest has probably been aroused by the questions of how seriously one can speak about this imaginary four-dimensional space, of the extent to which one can construct the geometry of this space by analogy with ordinary geometry, and of the differences and similarities between four-dimensional and three-dimensional geometry. Mathematicians who have studied these questions have obtained the following answer:

Yes, it is possible to develop such a geometry; it is in many respects similar to ordinary geometry. More-over, this geometry contains ordinary geometry as a special case, exactly as solid geometry (geometry in space) contains plane geometry as a special case. But, of course, the geometry of four-dimensional space will also have quite essential differences from ordinary geometry. The fantasy-writer H. G. Wells has written a very interesting story based on the peculiarities of a four-dimensional world.

But we will now show that these peculiarities are essentially quite similar to the peculiarities that

---

[1]We shall not study the derivation of the formulas for comput-ing the volume of the four-dimensional sphere. Here we shall mention however, that the volume of the four-dimensional sphere is equal to $\pi^2 R^4/2$. For comparison we point out that the volume of the five-dimensional sphere is equal to $8\pi^2 R^5/15$, that of the six-dimensional sphere is $\pi^3 R^6/6$, and that of the seven-dimensional one is $16\pi^3 R^7/105$.

distinguish the geometry of three-dimensional space from the geometry of the two-dimensional plane.

4. The Peculiarities of Four-Dimensional Space

Draw a circle in the plane and imagine yourself to be a creature in a two-dimensional world — or better, a point that can move on the plane but cannot go out into space. (You do not even know that space exists and cannot conceive of it.) Then the boundary of the circle, the circumference, will be an insurmountable barrier for you: you will not be able to leave the circle because the edge will block your path in every direction (Fig. 30*a*).

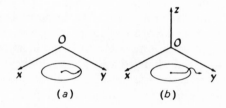

(*a*)   (*b*)

Fig. 30(*a*) The point, remaining within the limits of the plane, cannot leave the circle; (*b*) the point is free to leave the circle by going out into space.

Now imagine that this plane with the circle drawn in it is placed in three-dimensional space and that you have surmised the existence of a third dimension. You can now leave the limits of the circle without difficulty, of course, by simply stepping across the edge (Fig. 30*b*).

Now suppose you are a creature in a three-dimensional world (as before, if you do not object, we will consider you to be a point — this is, of course, entirely inessential). Suppose that you are situated inside a sphere beyond whose surface you cannot pass. You will be unable to leave the limits of this sphere. But if the sphere is placed in four-dimensional space and you have knowledge of the existence of a fourth dimension, then you will be able to leave the confines of the sphere without any difficulty.

There is nothing especially mystical about this — it is simply that the surface of the three-dimensional sphere does not separate four-dimensional space into two parts, although it does separate three-dimensional space. This is fully analogous to the fact that the boundary of a circle (the circumference) does not separate three-dimensional space, although it does separate the plane in which it lies.

One more example: It is clear that two figures in the plane which are mirror images of one another cannot be made to coincide without moving one of them out of the plane in which they lie. But a butterfly at rest unfolds its wings by moving them from the horizontal plane to the vertical (see the diagram on the back cover). Similarly, in a space of three dimensions it is impossible to make symmetric space figures coincide. For example, it is impossible to make a left-handed glove into a right-handed one although they are the same geometric shapes. But in a space of four-dimensions, three-dimensional symmetric figures can be made to coincide exactly as plane symmetric figures can be made to coincide if one moves them into three-dimensional space.

Thus, there is nothing surprising in the fact that the hero of the H. G. Wells story turned out to be reversed after his journey in four-dimensional space (his heart, for example, was now on the right, and his body was symmetric to what it had been before). This happened because when he went into four-dimensional space he was turned about in it.

5. Some Physics

Four-dimensional geometry has turned out to be an exceedingly useful and even an indispensable tool for modern physics. Without the tool of multi-dimensional geometry it would have been quite difficult to expound and use such important branches of contemporary physics as Einstein's theory of relativity.

Every mathematician can envy Minkowski, who, after using geometry very successfully in the theory of numbers, was able again with the aid of graphic geometric concepts to bring clarity to difficult mathematical questions — this time, concerning the theory of relativity. At the heart of the theory of relativity lies the idea of the indissoluble connection between space and time. That is, it is natural to consider the moment of time in which an event occurs as the fourth coordinate of this event together with the first three defining the point of space in which the event occurs.

The four-dimensional space so obtained is called the Minkowski space. A modern course in the theory of relativity will always begin with the description of this space. Minkowski's discovery was the fact that the principal formulas of the theory of relativity — the formulas of Lorentz — are quite simple when written in the terminology of the coordinates of this special four-dimensional space.

In this way, it was a great stroke of luck for modern physics that at the time of the discovery of the theory of relativity mathematicians had prepared the convenient, compact, and beautiful tool of multidimensional geometry, which in a number of cases significantly simplifies the solution of problems.

CHAPTER 2

Four-Dimensional Space

In the concluding chapters we shall discuss the geometry of four-dimensional space, as we promised earlier.

In the construction of geometry on the line, in the plane, and in three-dimensional space we have two possibilities: we can present the material with the aid of visual representations (since this is the method generally used in the school course, it is difficult to imagine a geometry textbook without sketches); or — and this is the possibility that the method of co-ordinates gives us — we can present it purely analytically, defining a point of the plane, for example, as a pair of numbers (the coordinates of the point), and a point in space as a triple of numbers.

For four-dimensional space the first possibility is not present. We cannot use visual geometric representations directly because the space surrounding us has three dimensions in all. The second way, however, is not barred to us. Indeed, we define a point of a line as a number, a point of a plane as a pair of numbers, and a point of three-dimensional space as a triple of numbers. Therefore it is completely natural to construct the geometry of four-dimensional space by defining a point of this imaginary space as a quadruple of numbers. By geometric figures in such a space we shall have to mean sets of points (just as in ordinary geometry). Let us proceed now to the exact definitions.

6. Coordinate Axes and Planes

Definition. An ordered[1] quadruple of numbers (x, y, z, u) is a *point* of four-dimensional space.

What are the coordinate axes in a space of four-dimensions and how many of them are there?

In order to answer this question, we return temporarily to the plane and three-dimensional space.

In the plane (that is, in a space of two dimensions) the coordinate axes are the sets of points one of whose coordinates can have any numerical value but whose other coordinate is equal to zero. Thus, the abscissa axis is the set of points of the form $(x, 0)$, where x is any number. For example, the points $(1, 0)$, $(-3, 0)$, $(2\frac{1}{3}, 0)$ all lie on the abscissa axis; but the point $(\frac{1}{5}, 2)$ does not lie on this axis. Similarly, the ordinate axis is the set of points of the form $(0, y)$, where y is any number.

Three-dimensional space has three axes:

The x-axis — the set of points of the form $(x, 0, 0)$, where x is any number.

The y-axis — the set of points of the form $(0, y, 0)$, where y is any number.

The z-axis — the set of points of the form $(0, 0, z)$, where z is any number.

In four-dimensional space consisting of all points of the form (x, y, z, u), where x, y, z, and u are arbitrary numbers, it is natural to take the *coordinate axes* to be the sets of points one of whose coordinates can take on arbitrary numerical values but whose

[1]We say "ordered," since different orderings of the same numbers in a quadruple give different points: for example, the point $(1, -2, 3, 8)$ is different from the point $(3, 1, 8, -2)$.

remaining coordinates are equal to zero. Then it is clear that four-dimensional space has four coordinate axes:

The x-axis — the set of points of the form $(x, 0, 0, 0)$, where x is any number.

The y-axis — the set of points of the form $(0, y, 0, 0)$, where y is any number.

The z-axis — the set of points of the form $(0, 0, z, 0)$, where z is any number.

The u-axis — the set of points of the form $(0, 0, 0, u)$, where u is any number.

In three dimensional space there are, in addition to the coordinate axes, the *coordinate planes*. These are the planes passing through any pair of coordinate axes. The yz-plane, for example, is the plane passing through the y- and z-axes. In three-dimensional space there are three coordinate planes in all:

The xy-plane — the set of points of the form $(x, y, 0)$, where x and y are arbitrary numbers.

The yz-plane — the set of points of the form $(0, y, z)$, where y and z are arbitrary numbers.

The xy-plane — the set of points of the form $(x, 0, z)$, where x and z are arbitrary numbers.

Thus it is natural to define the *coordinate planes* in four-dimensional space as the sets of points for which two of the four coordinates take on arbitrary numerical values and the other two are equal to zero. For example, we shall take as the xz coordinate plane in four-dimensional space the set of points of the form $(x, 0, z, 0)$. How many of these planes are there in all?

This is not difficult to figure out. We can simply write them all down:

The xy-plane — the set of points of the form $(x, y, 0, 0)$.

The xz-plane — the set of points of the form $(x, 0, z, 0)$.

The *xu*-plane — the set of points of the form $(x, 0, 0, u)$.

The *yz*-plane — the set of points of the form $(0, y, z, 0)$.

The *yu*-plane — the set of points of the form $(0, y, 0, u)$.

The *zu*-plane — the set of points of the form $(0, 0, z, u)$.

For each of these planes the variable coordinates can take on arbitrary numerical values, including zero. For example, the point $(5, 0, 0, 0)$ belongs to the *xy*-plane and to the *xu*-plane (and to which other?). Thus it is easy to see that the *yz*-plane, for example, "passes" through the *y*-axis in the sense that each point of the *y*-axis belongs to this plane. For in fact, any point on the *y*-axis — that is, any point of the form $(0, y, 0, 0)$ — belongs to the set of points of the form $(0, y, z, 0)$ — that is, to the *yz*-plane.

Question. What set is formed by the points belonging simultaneously to the *yz*-plane and to the *xz*-plane?

Answer. This set consists of all points of the form $(0, 0, z, 0)$ — that is, it is the *z*-axis.

Thus, there exist in four-dimensional space sets of points analogous to the coordinate planes of three-dimensional space. There are six of them. Each of them consists of the points that, like the points of the coordinate planes of three-dimensional space, have two coordinates allowed to take on arbitrary numerical values and have the remaining coordinates equal to zero. Each of these coordinate planes "passes" through two coordinate axes: the *yz*-plane, for example, passes through the *y*-axis and the *z*-axis. On the other hand, three coordinate planes pass through each axis. For example, the *xy*-, *xz*-, and *xu*-planes pass through the *x*-axis. We shall therefore say that the *x*-axis is the intersection of these planes.

The six coordinate axes contain only one point in common. This is the point $(0, 0, 0, 0)$ — the origin.

Question. What set of points is the intersection of the *xy*-plane and the *yz*-plane? of the *xy*-plane and the *zu*-plane?

We see that we obtain a picture fully analogous to the one in three-dimensional space. We can even try to make a schematic diagram that will help to create some visual model for the disposition of the coordinate planes and axes of four-dimensional space. In Fig. 31 the coordinate planes are depicted by parallelograms, and the axes, by straight lines: everything is exactly as in Fig. 20 for three-dimensional space.

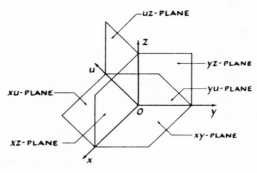

Fig. 31

There are, however, still other sets of points in four-dimensional space which can be called coordinate planes. One should, incidentally, have expected this: for the straight line has only the origin; the plane has both the origin and the axes; and three-dimensional space has the coordinate planes in addition to the origin and the axes. Thus it is natural that in four-dimensional space new sets should appear, which we shall call the *three-dimensional coordinate planes*.

These planes are the sets consisting of all points for which three of the four coordinates take on all possible numerical values but the fourth coordinate is equal to zero. An example of one of these three-dimensional coordinate planes is the set of points of the form $(x, 0, z, u)$, where x, z, and u take on all possible values. This set is called the *three-dimensional coordinate plane xzu*. It is easy to see that in four-dimensional space there exist four three-dimensional coordinate planes:

The *xyz*-plane — the set of points of the form $(x, y, z, 0)$.

The *xyu*-plane — the set of points of the form $(x, y, 0, u)$.

The *xzu*-plane — the set of points of the form $(x, 0, z, u)$.

The *yzu*-plane — the set of points of the form $(0, y, z, u)$.

One can say, too, that each of the three-dimensional coordinate planes "passes" through the origin and that each of these planes "passes" through three of the coordinate axes (we use the word "passes" here in the sense that the origin and each of the points of the axes belong to the plane). For example, the three-dimensional plane *xyu* passes through the axes x, y, and u.

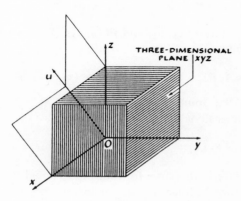

Fig. 32

Analogously, one can say that each of the two-dimensional planes is the intersection of two three-dimensional planes. The xy-plane, for example, is the intersection of the xyz-plane and the xyu-plane, that is, consists of all points belonging simultaneously to each of these three-dimensional planes.

Examine Fig. 32. It is different from Fig. 31 in that we have drawn in it the three-dimensional coordinate plane xyz. It is depicted as a parallelepiped. It is evident that this plane contains the x-, y-, and z-axes and the xy-, xz-, and yz-planes.

7. Some Problems

Let us now try to determine in what sense we can speak of the distance between points of four-dimensional space.

In Sections 3, 6, and 9 of Part 1 of this volume we showed that the method of coordinates enables us to define the distance between points without relying upon a geometric representation. In fact, the distance can be computed for the points $A(x_1)$ and $B(x_2)$ of the line by the formula

$$d(A, B) = |x_1 - x_2|,$$

or

$$d(A, B) = \sqrt{(x_1 - x_2)^2};$$

for the points $A(x_1, y_1)$ and $B(x_2, y_2)$ of the plane by the formula

$$d(A, B) = \sqrt{(x_1 - x_2)^2 + (y_1 - y_2)^2},$$

and for the points $A(x_1, y_1, z_1)$ and $B(x_2, y_2, z_2)$ of three-dimensional space by the formula

$$d(A, B) = \sqrt{(x_1 - x_2)^2 + (y_1 - y_2)^2 + (z_1 - z_2)^2}.$$

It is natural therefore for four-dimensional space to define the distance in an analogous way and to introduce the following

60

Definition. The *distance* between two points $A(x_1, y_1, z_1, u_1)$ and $B(x_2, y_2, z_2, u_2)$ of four-dimensional space is defined to be the number $d(A, B)$ given by the formula

$$d(A, B) = \sqrt{(x_1 - x_2)^2 + (y_1 - y_2)^2 + (z_1 - z_2)^2 + (u_1 - u_2)^2}.$$

In particular, the distance of the point $A(x, y, z, u)$ from the origin is given by the formula

$$d(O, A) = \sqrt{x^2 + y^2 + z^2 + u^2}.$$

Using this definition, one can solve problems of the geometry of four-dimensional space quite like those that one solves in school problem-books.

EXERCISES

1. Prove that the triangle with vertices $A(4, 7, -3, 5)$, $B(3, 0, -3, 1)$, and $C(-1, 7, -3, 0)$ is isosceles.

2. Consider the four points of four-dimensional space: $A(1, 1, 1, 1)$, $B(-1, -1, 1, 1)$, $C(-1, 1, 1, -1)$, and $D(1, -1, 1, -1)$. Prove that these four points are equidistant from one another.

3. Let A, B, and C be points of four-dimensional space. We can define the angle ABC in the following way. As we are able to compute distances in four-dimensional space, we can find $d(A, B)$, $d(B, C)$, and $d(A, C)$, that is, the "lengths of the sides" of the triangle ABC. We now construct in the ordinary two-dimensional plane a triangle $A'B'C'$ such that its sides $A'B'$, $B'C'$, and $C'A'$ are equal to $d(A, B)$, $d(B, C)$, and $d(C, A)$, respectively. Then the angle $A'B'C'$ of this triangle is defined to be the *angle ABC* in four-dimensional space.

Prove that the triangle with vertices $A(4, 7, -3, 5)$, $B(3, 0, -3, 1)$, and $C(1, 3, -2, 0)$ is a right triangle.

4. Take the points A, B, and C of Exercise 1. Compute the angles A, B, and C of the triangle ABC.

CHAPTER 3

The Four-Dimensional Cube

8. The Definition of the Sphere and the Cube

Let us now consider geometric figures in four-dimensional space. By a geometric figure (as in ordinary geometry) we shall mean some set of points.

Let us consider the definition of a sphere: a sphere is the set of points whose distance from some fixed point is a certain fixed value. This definition can be used to define a sphere in four-dimensional space, for we know what a point in four-dimensional space is, and we also know what the distance between two points is. We thus take the same definition, translating it into the terminology of numbers (for simplicity, as in the case of three-dimensional space, we take the center of the sphere to be the origin).

Definition. The set of points (x, y, z, u) satisfying the relation

$$x^2 + y^2 + z^2 + u^2 = R^2$$

is called the *four-dimensional sphere* with center at the origin and radius R.

Let us now discuss the four-dimensional cube. From the name, we see that this is a figure analogous to the familiar three-dimensional cube (Fig. 33a). In the plane there is also a figure analogous to the cube — the square. One can see the analogy between them particularly easily if one examines the analytic definitions of the cube and the square.

In fact (as you already know from Exercise 4 in Section 10 of Part 1), one can give the following definition:

The *cube* is the set of points (x, y, z) satisfying the relations

$$0 \le x \le 1,$$
$$0 \le y \le 1, \qquad (*)$$
$$0 \le z \le 1.$$

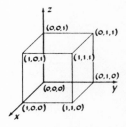

Fig. 33a

This "arithmetical" definition does not require any sketch. But it fully corresponds to the geometric definition of the cube.[1]

For the square one can also give an arithmetical definition:

The *square* is the set of points (x, y) satisfying the relations (Fig. 33b)

$$0 \le x \le 1,$$
$$0 \le y \le 1.$$

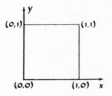

Fig. 33b

Comparing these two definitions, one easily sees that the square really is, as they say, the two-dimensional analogue of the cube. We shall sometimes call the square the "two-dimensional cube."

One can also examine an analogue of these figures in a space of one dimension, that is, on the line. For we can take the set of points x of the line satisfying

---

[1] Of course, there are other cubes in space as well. For example, the set of points defined by the relations $-1 \le x \le 1$, $-1 \le y \le 1$, $-1 \le z \le 1$ is also a cube. This cube is quite conveniently situated with respect to the coordinate axes: the origin is its center; and the coordinate axes and planes are the axes and planes of symmetry. However, we have decided to consider as fundamental the cube defined by relations (*). We shall sometimes call this cube the unit cube in order to distinguish it from other cubes.

the relations

$$0 \le x \le 1.$$

It is clear that this "one-dimensional cube" is a line segment (Fig. 33c).

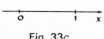

Fig. 33c

Hopefully, then, you will now accept as completely natural the following

Definition. The *four-dimensional cube* is the set of points (x, y, z, u) satisfying the relations

$$0 \le x \le 1,$$
$$0 \le y \le 1,$$
$$0 \le z \le 1,$$
$$0 \le u \le 1.$$

There is no need to be distressed because we have not introduced a picture of the four-dimensional cube; we shall do this later on. (Do not be surprised that it is possible to draw the four-dimensional cube: after all, we draw the three-dimensional cube on a flat sheet of paper.) In order to give a drawing of the four-dimensional cube, however, it will be necessary first to discuss how this cube is "constructed" and what elements in it can be distinguished.

9. The Structure of the Four-Dimensional Cube

Let us examine the "cubes" of various dimensions, that is, the line segment, the square, and the ordinary cube.

The segment, defined by relations $0 \le x \le 1$ is a very simple figure. All we can say about it, maybe, is that its boundary consists of two points: 0 and 1. The remaining points of the segment we shall call interior points.

The boundary of the square consists of four points (the vertices) and four segments. Thus, the square has on its boundary elements of two types: points and line segments. The boundary of the three-dimensional cube contains elements of three types: vertices — eight of them, edges (line segments) — twelve of them, and faces (squares) — of which there are six.

Let us write this down in a table. This table can be abbreviated if we agree to write, instead of the name

Composition of the Boundary (The Figure)	Points (Vertices)	Segments (Sides, Edges)	Squares (Faces)
The Segment . .	2	—	—
The Square . . .	4	4	—
The Cube	8	12	6

of the figure, the number n equal to its dimension: for the segment, $n = 1$; for the square, $n = 2$; for the cube, $n = 3$. Instead of the name of the element of the boundary, we can likewise write down merely the dimension of this element: for the face, $n = 2$; for the edge, $n = 1$. For convenience, we consider the point (the vertex) to have zero dimension ($n = 0$). Then this table takes on a different form.

Dimension of the Boundary	0	1	2
Dimension of the Cube			
1	2	—	—
2	4	4	—
3	8	12	6
4			

Our aim is to complete the fourth row of this table. For this, we once again examine the boundaries of the segment, the square, and the cube, but this time analytically,[1] and try to see by analogy how the boundary of the four-dimensional cube is constructed.

[1]That is, purely arithmetically.

The boundary of the segment $0 \leq x \leq 1$ consists of two points: $x = 0$ and $x = 1$.

The boundary of the square $0 \leq x \leq 1, 0 \leq y \leq 1$ contains four vertices: $x = 0, y = 0$; $x = 0, y = 1$; $x = 1, y = 0$; and $x = 1, y = 1$, that is, the points $(0, 0)$, $(0, 1)$, $(1, 0)$, and $(1, 1)$.

The cube $0 \leq x \leq 1$, $0 \leq y \leq 1$, $0 \leq z \leq 1$ contains eight vertices. Each of these is a point (x, y, z) in which x, y, and z are either 0 or 1. One obtains the following eight points: $(0, 0, 0)$, $(0, 0, 1)$, $(0, 1, 0)$, $(0, 1, 1)$, $(1, 0, 0)$, $(1, 0, 1)$, $(1, 1, 0)$, $(1, 1, 1)$.

The *vertices* of the four-dimensional cube,

$$0 \leq x \leq 1,$$
$$0 \leq y \leq 1,$$
$$0 \leq z \leq 1,$$
$$0 \leq u \leq 1,$$

are taken to be the points (x, y, z, u) for which x, y, z, and u are either 0 or 1.

There are sixteen such vertices, for it is possible to write down sixteen different quadruples of zeros and ones. In fact, let us take the triples composed of the coordinates of the vertices of the three-dimensional cube (there are eight of them), and to each such triple let us assign first 0, then 1. Then in this way, for each such triple we get two quadruples; and so there will be $8 \times 2 = 16$ quadruples in all. Thus we have computed the number of vertices of the four-dimensional cube.

Let us consider now what we should call the edge of the four-dimensional cube. Again we make use of analogy. For the square the edges (sides) are defined by the following relations (see Fig. 33*b*):

$$0 \leq x \leq 1, \qquad y = 0 \text{ (edge } AB\text{)};$$
$$x = 1, \qquad 0 \leq y \leq 1 \text{ (edge } BC\text{)};$$
$$0 \leq x \leq 1, \qquad y = 1 \text{ (edge } CD\text{)};$$
$$x = 0, \qquad 0 \leq y \leq 1 \text{ (edge } DA\text{)}.$$

As we see, the edges of the square are characterized by the property that for each point of a given edge, one of the coordinates has a definite numerical value: 0 or 1, whereas the second coordinate can take on all values between 0 and 1.

Let us further examine the edges of the (three-dimensional) cube. We have (see Fig. 33a)

$$x = 0, \qquad y = 0, \quad 0 \le z \le 1 \quad \text{(edge } AA_1\text{)};$$
$$0 \le x \le 1, \qquad y = 0, \qquad z = 1 \quad \text{(edge } A_1B_1\text{)};$$
$$x = 1, \qquad 0 \le y \le 1, \qquad z = 1 \quad \text{(edge } B_1C_1\text{)},$$

and so on.

By analogy we give the following

Definition. The *edges* of the four-dimensional cube are the sets of points for which all of the coordinates except one are constant (and equal to 0 or 1), whereas the fourth can take on all possible values from 0 to 1.

Examples of edges are

(1) $x = 0, \qquad y = 0, \quad z = 1, \ 0 \le u \le 1;$

(2) $0 \le x \le 1, \qquad y = 1, \quad z = 0, \qquad u = 1;$

(3) $x = 1, \qquad 0 \le y \le 1, \quad z = 0, \qquad u = 0,$

and so on.

Let us try to compute the number of edges of the four-dimensional cube, that is, the number of such lines that can be written down. In order to avoid becoming confused we shall count them in a definite order. First, we shall distinguish four groups of edges: for the first group let the variable coordinate be $x(0 \le x \le 1)$, and let y, z, and u have the constant values 0 and 1 in all possible combinations. We already know that there exist 8 different triples consisting of zeros and ones (recall how many vertices the three-dimensional cube has). Therefore there exist 8 edges of the first group (for which the variable coordinate is x). It is easy to see that there are likewise

8 edges in the second group, for which the variable coordinate is not x but y. Thus it is clear that the four-dimensional cube has $4 \times 8 = 32$ edges.

We can now easily write down the relations defining each of these edges without fear of leaving out any of them:

First Group:

$0 \le x \le 1$

y	z	u
0	0	0
0	0	1
0	1	0
0	1	1
1	0	0
1	0	1
1	1	0
1	1	1

Second Group:

$0 \le y \le 1$

x	z	u
0	0	0
0	0	1
0	1	0
...	...	...

Third Group:

$0 \le z \le 1$

x	y	u
0	0	0
0	0	1
...	...	...

Fourth Group:

$0 \le u \le 1$

x	y	z
0	0	0
0	0	1
...	...	...

The three-dimensional cube has faces, in addition to vertices and edges. On each of the faces two coordinates vary (taking on all possible values from 0 to 1), but one coordinate is constant (equal to 0 or 1). For example, the face ABB_1A (Fig. 33a) is defined by the relations

$$0 \le x \le 1, \qquad y = 0, \qquad 0 \le z \le 1.$$

By analogy we can give the following

Definition. A *two-dimensional face*[1] of the four-dimensional cube is the set of points for which any two coordinates can take on all possible values between 0 and 1, whereas the other two remain constant (equal to either 0 or 1).

Examples of faces are the following:

$$x = 0, \quad 0 \leq y \leq 1, \quad z = 1, \quad 0 \leq u \leq 1.$$

EXERCISE

Calculate the number of faces of the four-dimensional cube. (**Hint.** We advise you not to resort to a sketch but to use only analytic (arithmetical) definitions and to write down all six rows of relations defining each of the six faces of the ordinary three-dimensional cube. **Answer.** The four-dimensional cube has 24 two-dimensional faces.)

We can now fill in the fourth row of our table. The table, however, is clearly still incomplete: the entry

Dimension of the Boundary	0	1	2	3
Dimension of the Cube				
1	2	—	—	—
2	4	4	—	—
3	8	12	6	—
4	16	32	24	

in the lower right-hand corner is missing. The fact is that for the four-dimensional cube it will be necessary to add another column. For the segment, indeed, there was only one type of boundary: the vertices; the square had two types: vertices and edges; and the cube had two-dimensional faces as well. One should expect, therefore, that the four-dimensional cube will

---

[1]The necessity for specifying that the face should be two-dimensional will be explained somewhat later.

have a new type of element making up its boundary in addition to those we have seen and that the dimension of this new element will be equal to three.

We therefore give the following

Definition. A *three-dimensional face* of the four-dimensional cube is a set of points for which three of the coordinates take on all possible values from 0 to 1 and the fourth is constant (equal to 0 or to 1).

One can easily compute the number of three-dimensional faces. There are eight of them, since for each of the four coordinates there are two possible values: 0 and 1, and we have $2 \times 4 = 8$.

Let us now look at Fig. 34. Here we have drawn a four-dimensional cube. All 16 vertices are visible in the diagram, as well as the 32 edges, the 24 two-dimensional faces (shown as parallelograms), and the 8 three-dimensional faces (shown as parallelepipeds). From the diagram it is quite clear which face contains which edge, and so on.

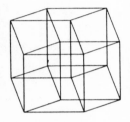

Fig. 34

How is this diagram obtained? Consider how one draws the ordinary cube on a flat sheet of paper. One really draws the so-called parallel projection of the three-dimensional cube on the two-dimensional plane.[1] In order to obtain our diagram, we first make a space model of the projection of the four-dimensional cube onto three-dimensional space and then draw this

[1] In a course in solid geometry you will become more familiar with the parallel projection. In order to see what the parallel projections of the ordinary cube on the plane are, proceed in this way: make a cube out of wire (that is, make the framework of a cube) and examine the shadow that it casts on a sheet of paper or on a wall on a sunny day. If you place the cube properly, the shadow you obtain will be the figure that you usually see in books. This is the parallel projection of the cube onto the plane. To obtain it, one must construct a straight line through each point of the cube parallel to a fixed direction (the sun's rays are all parallel to one another) but not necessarily perpendicular to the plane. Then the intersection of these lines with the plane onto which we are projecting is the parallel projection of the figure.

model. If you are skillful, then you too can make such a model. You can, for example, use ordinary matches, fastening them together with wax beads. (How many matches will you need? And how many beads? How many matches must be inserted in each bead?)

One can obtain a visual representation of the four-dimensional cube by other means as well. Suppose that we have asked you to send us a model of the ordinary three-dimensional cube. You could, of course, mail a "three-dimensional" package, but this is involved. Therefore it is better to proceed as follows: glue the cube together out of paper; then unfasten the cube and send us the pattern or, as mathematicians would say, the development of the cube. Such a development is depicted in Fig. 35. As the coordinates of the vertices have been inserted in the figure, one can easily see how to fasten together the pattern in order to obtain the cube itself.

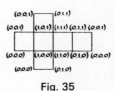

Fig. 35

EXERCISES

1. Write down the relations defining each of the three-dimensional faces of the four-dimensional cube.

2. One can construct a development of the four-dimensional cube. This will be a three-dimensional figure. It is evident that it will consist of 8 cubes. If you succeed in making the development or in seeing how it should be made, make a drawing of it and show the coordinates of each vertex on the drawing.

10. Problems on the Cube

We have discussed the construction of the four-dimensional cube. Let us now talk about its dimensions. The length of each of the edges of the four-dimensional cube is equal to one, just as in the ordinary cube and the square (by the length of an edge we mean

71

the distance between the vertices lying on this edge). For this reason we have called our "cubes" unit cubes.

1. Calculate the distances between the vertices of the cube not lying on a single edge. (Take one of the vertices, say $(0, 0, 0, 0)$, and calculate the distance between this vertex and the others. You have the formula for computing the distance between points; and since you know the coordinates of the vertices, all that remains is to carry out some simple computations.)

2. Having solved Problem 1, you see that the vertices can be classified into four groups. The vertices of the first group are located at a distance of 1 from $(0, 0, 0, 0)$; the vertices of the second group are a distance of $\sqrt{2}$ from this point; the vertices of the third group are $\sqrt{3}$ away; and those of the fourth are $\sqrt{4} = 2$ away. How many of the vertices of the four-dimensional cube are in each group?

3. The vertex $(1, 1, 1, 1)$ is located at the greatest distance from the vertex $(0, 0, 0, 0)$; that is, its distance from this point is equal to 2. We shall call this vertex the vertex *opposite* the vertex $(0, 0, 0, 0)$; the segment joining them is called the *main diagonal* of the four-dimensional cube. What should one take to be the main diagonal for cubes of other dimensions, and what are the lengths of their main diagonals?

4. Suppose now that the three-dimensional cube is made of wire and that an ant is sitting at the vertex $(0, 0, 0)$. Suppose further that the ant must crawl from one vertex to the other. How many edges must the ant cross in order to get from the vertex $(0, 0, 0)$ to the vertex $(1, 1, 1)$? It must cross three edges. Therefore we shall call the vertex $(1, 1, 1)$ a vertex of the third order. The path from the vertex $(0, 0, 0)$ to the vertex $(0, 1, 1)$ along the edges consists of two links. Such a vertex we shall call a vertex of the second order. In the cube there are vertices of the first order as well: those which the ant can get to by traversing a single edge. There are three such vertices: $(0, 0, 1)$, $(0, 1, 0)$, and $(1, 0, 0)$. The cube also has three vertices of

the second order. Write down their coordinates (Problem 4*a*). There exist two paths from (0, 0, 0) to each of the vertices of the second order consisting of two links. For example, one can get to the vertex (0, 1, 1) through the vertex (0, 0, 1) and also through the vertex (0, 1, 0). How many paths containing three links are there connecting a vertex with its opposite vertex (Problem 4*b*)?

5. Take the four-dimensional cube with the center at the origin, that is, the set of points satisfying the following relations:

$$-1 \leq x \leq 1,$$
$$-1 \leq y \leq 1,$$
$$-1 \leq z \leq 1,$$
$$-1 \leq u \leq 1.$$

Find the distance from the vertex (1, 1, 1, 1) to each of the other vertices of this cube.

Which vertices will be vertices of the first order with respect to the vertex (1, 1, 1, 1) (that is, which vertices can one get to from the vertex (1, 1, 1, 1), traversing only one edge)? Which vertices will be vertices of the second order? of the third? of the fourth?

6. And the last question, to test your understanding of the four-dimensional cube: How many paths are there having four links and leading from the vertex (0, 0, 0, 0) of the four-dimensional cube to the opposite vertex (1, 1, 1, 1) going along the edges of this cube? Write down each path specifically, showing in order the vertices that one must pass.

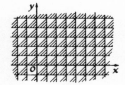

Answer to Exercise 1*f* on page 18.